PRINCIPLES AND TECHNIQUES OF PLANT BREEDING

Edited by

Razel Paniza Almendras

ArclerPress

Principles and Techniques of Plant Breeding

Edited by Razel Paniza Almendras

Edition 2018

ArclerPress

Published by
Arcler Press LLC
708 3rd Avenue, 6th Floor
New York
NY 10017
United States of America
www.arclerpress.com

ISBN: 978-1-68094-560-7 (Hardback)

Printed in United States of America

Library of Congress Control Number: 2017933130

About the Editor

Razel Paniza Almendras

Razel completed her Bachelor's Degree in Agriculture (major in Crop Science) at Xavier University-Ateneo de Cagayan, Philippines last 2013. She passed the licensure examination for agriculturist at the same year. Her interest involves researches developing protocols in tissue cultured commodities. Now, she is currently working as Date-Palm Tissue Culturist in United Arab of Emirates (UAE).

Table of Contents

3 MOLECULAR BREEDING

4 SELECTION METHODS IN PLANT BREEDING BASED ON MODE OF REPRODUCTION

5 SMART BREEDING

6 BREEDING IN SELF-POLLINATION PLANTS

7 BREEDING FOR DROUGHT STRESS TOLERANCE

8 ARTIFICIAL PRODUCTION OF DOUBLED HAPLOIDS IN PLANT BREEDING

List of Abbreviations

BSA	Bulked Segregant Analysis
CLL	Chronic Lymphocytic Leukemia
CMS	Cytoplasmic Male Sterility
CMS	Cell Membrane Stability
CTD	Canopy Temperature Depression
DH	Doubled Haploid
DSB	Double-Stranded Breaks
HSP	Heat Shock Proteins
ICNCP	International Code of Nomenclature for Cultivated Plants
ICRA	International Cultivar Registration Authority
IVDMD	In Vitro Dry Matter Digestibility
MAB	Marker-Assisted Breeding
MABC	Marker-Assisted Backcrossing
MAS	Marker Aided Selection
MB	Molecular Breeding
OA	Osmotic Adjustment
QTL	Quantitative Trait Locus
RHS	Royal Horticultural Society
RIL	Recombinant Inbred Lines
ROS	Reactive Oxygen Species
SNP	Single Nucleotide Polymorphisms

Preface

The world's population is growing. Demand for food and water is increasing. Over the next few decades, food production will need to increase dramatically to keep pace with population growth and meet the needs of changing diets. To do that, farmers will need to grow as much food in the next 50 years as they did in the past 10,000 years of farming combined, while managing with a declining amount of resources. To successfully accomplish these goals, innovations in plant sciences must be at the forefront of what we do. Advances in plant breeding are helping increase the rate of improvement in key plant characteristics, such as grain yield and disease tolerance. Plant breeding is the process by which humans change the characteristics of plants over time to make them better crops and more nourishing food. In its most simple form, breeding consists of selecting the best plants in a given field, growing them to full seed and then using that seed to grow further generations. Plant breeding can be accomplished through many different techniques ranging from simply selecting plants with desirable characteristics for propagation, to more complex molecular techniques. Plant breeding has been practiced for thousands of years, since near the beginning of human civilization. It is now practiced worldwide by individuals such as gardeners and farmers, or by professional plant breeders employed by organizations such as government institutions, universities, crop-specific industry associations or research centers. Plant breeding in certain situations may lead the domestication of wild plants. Domestication of plants is an artificial selection process conducted by humans to produce plants that have more desirable traits than wild plants, and which renders them dependent on artificial (usually enhanced) environments for their continued existence. A plant whose origin or selection is due primarily to intentional human activity is called a cultigen, and a cultivated crop species that has evolved from wild populations due to selective pressures from traditional farmers is called a landrace. Landraces, which can be the result of natural forces or domestication, are plants (or animals) that are ideally suited to a particular region or environment. An example are the landraces of rice, *Oryza sativa* subspecies indica, which was developed in South Asia, and *Oryza sativa* subspecies *japonica*, which was developed in China. Consequently, Plant breeding is the purposeful manipulation of plant species in order to create desired genotypes and phenotypes for specific purposes.

Content Coverage

Chapter One is intended to present an overview of plant breeding practiced worldwide by individuals such as gardeners and farmers, or by professional plant breeders employed by organizations such as government institutions, universities, crop-specific industry associations or research centers.

Chapter Two provides the information about most commonly cultivars, referring to an assemblage of plants selected for desirable characteristics that are maintained during propagation. More generally, cultivar refers to the most basic classification category of cultivated plants. Most cultivars have arisen in cultivation, but a few are special selections from the wild.

Chapter Three describes several molecular breeding strategies as the use of genetic manipulation performed at DNA molecular levels to improve characters of interest in plants and animals, including genetic engineering or gene manipulation, molecular marker-assisted selection, genomic selection, etc.

Chapter Four outlines the selection methods in plant breeding based on mode of reproduction. The mode of reproduction of a crop determines its genetic composition, which, in turn, is the deciding factor to develop suitable breeding and selection methods. Knowledge of mode of reproduction is also essential for its artificial manipulation to breed improved types.

Chapter Five focuses on smart breeding methods that enables the detection of desired performance features irrespective of environmental influences. The analysis of the trait in question is a quick and easily automated process allowing a high throughput.

Chapter Six presents a framework on breeding in self-pollinated plants. For the process of pollination to be successful, a pollen grain produced by the anther, the male part of a flower, must be transferred to a stigma, the female part of the flower, of a plant of the same species.

Chapter Seven focuses on breeding for drought stress tolerance with the goal of reducing the impact of drought on plant growth.

Chapter Eight emphasizes on artificial production of doubled haploids in plant breeding. Dihaploid plants derived from tetraploid crop plants may be important for breeding programs that involve diploid wild relatives of the crops.

CHAPTER 1

An Overview of Plant Breeding

Chapter Contents

- **Introduction**
- **History**
- **Scientific Plant Breeding**
- **Classical Plant Breeding**
- **Modern Plant Breeding**
- **Role of Plant Breeding in Organic Agriculture**

Plant breeding is the art and science of changing the genetics of plants for the benefit of humankind. Plant breeding can be accomplished through many different techniques ranging from simply selecting plants with desirable characteristics for propagation, to more complex molecular techniques. Plant breeding has been practiced for thousands of years, since near the beginning of human civilization. It is now practiced worldwide by individuals such as gardeners and farmers, or by professional plant breeders employed by organizations such as government institutions, universities, crop-specific industry associations or research centers.

International development agencies believe that breeding new crops is important for ensuring food security by developing new varieties that are higher-yielding, resistant to pests and diseases, drought-resistant or regionally adapted to different environments and growing conditions.

INTRODUCTION

Plant breeding is the art and science of changing the traits of plants in order to produce desired characteristics. Plant breeding can be accomplished through many different techniques ranging from simply selecting plants with desirable characteristics for propagation, to more complex molecular techniques.

Plant breeding has been practiced for thousands of years, since near the beginning of human civilization. It is practiced worldwide by individuals such as gardeners and farmers, or by professional plant breeders employed by organizations such as government institutions, universities, crop-specific industry associations or research centers.

International development nation agencies believe that breeding new crops is important for ensuring food security by developing new varieties that are higher-yielding, disease resistant, drought-resistant or regionally adapted to different environments and growing conditions.

The Yecoro wheat (right) cultivar is sensitive to salinity, plants resulting from a hybrid cross with cultivar W4910 (left) show greater tolerance to high salinity.

HISTORY

Plant breeding started with sedentary agriculture and particularly the domestication of the first agricultural plants, a practice which is estimated to date back 9,000 to 11,000 years. Initially early farmers simply selected food plants with particular desirable characteristics, and employed these as progenitors for subsequent generations, resulting in an accumulation of valuable traits over time.

Gregor Mendel's experiments with plant hybridization led to his establishing laws of inheritance. Once this work became well known, it formed the basis of the new science of genetics, which stimulated research by many plant scientists dedicated to improving crop production through plant breeding.

Modern plant breeding is applied genetics, but its scientific basis is broader, covering molecular biology, cytology, systematics, physiology, pathology, entomology, chemistry, and statistics (biometrics). It has also developed its own technology.

Early Plant Breeding

Domestication

Domestication of plants is an artificial selection process conducted by humans to produce plants that have more desirable traits than wild plants, and which renders them dependent on artificial usually enhanced environments for their continued existence. The practice is estimated to date back 9,000-11,000 years. Many crops in present-day cultivation are the result of domestication in ancient times, about 5,000 years ago in the Old World and 3,000 years ago in the New World. In the Neolithic period, domestication took a minimum of 1,000 years and a maximum of 7,000 years. Today, all principal food crops come from domesticated varieties. Almost all the domesticated plants used today for food and agriculture were domesticated in the centers of origin. In these centers there is still a great diversity of closely related wild plants, so-called crop wild relatives, that can also be used for improving modern cultivars by plant breeding.

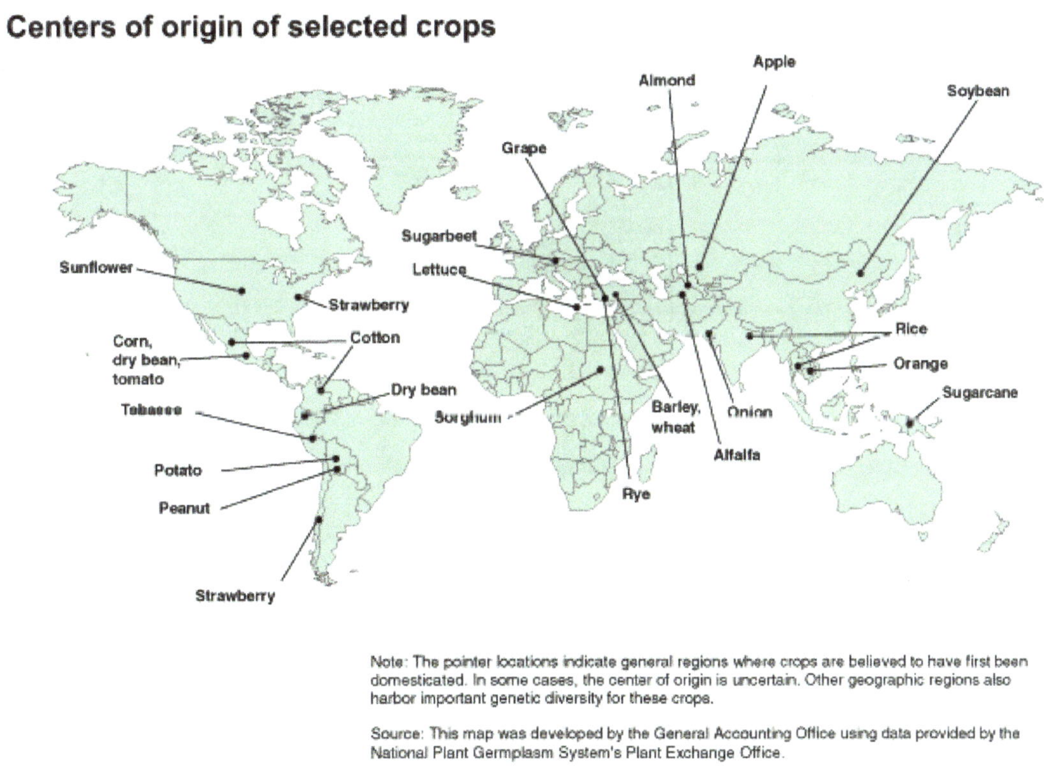

This map shows the sites of domestication for a number of crops. Places where crops were initially domesticated are called centers of origin.

A plant whose origin or selection is due primarily to intentional human activity is called a cultigen, and a cultivated crop species that has evolved from wild populations due to selective pressures from traditional farmers is called a landrace. Landraces, which can be the result of natural forces or domestication, are plants or animals that are ideally suited to a particular region or environment. An example are the landraces of rice, *Oryza sativa*subspecies *indica*, which was developed in South Asia, and *Oryza sativa* subspecies *japonica*, which was developed in China.

Columbian Exchange

Humans have traded useful plants from distant lands for centuries, and plant hunters have been sent to bring plants back for cultivation. Human agriculture has had two important results: the plants most favoured by humans came to be grown in many places and (2) gardens and farms have provided some opportunities for plants to interbreed that would not have been possible for their wild ancestors. Columbus's arrival in America in 1492 triggered unprecedented transfer of plant resources between Europe and the New World.

SCIENTIFIC PLANT BREEDING

Gregor Mendel's experiments with plant hybridization led to his laws of inheritance. This work became well known in the 1900s and formed the basis of the new science of genetics, which stimulated research by many plant scientists dedicated to improving crop production through plant breeding.

Garton's catalogue from 1902.

However, successful commercial plant breeding concerns began to be founded from the late 19th century. Gartons Agricultural Plant Breeders in England was established in the 1890s by John Garton, who was one of the first to cross-pollinate agricultural plants and commercialize the newly created varieties. He began experimenting with the artificial cross pollination firstly of cereal plants, then herbage species and root crops and developed far reaching techniques in plant breeding.

From 1904 to World War II in Italy, Nazareno Strampelli created a number of wheat hybrids. His work allowed Italy to increase crop production during the so-called "Battle for Grain" (1925–1940) and some varieties were exported to foreign countries, such as Argentina, Mexico, and China. Strampelli's work laid the foundations for Norman Borlaug and the Green Revolution.

Green Revolution

In 1908, George Harrison Shull described heterosis, also known as hybrid vigor. Heterosis describes the tendency of the progeny of a specific cross to outperform both parents. The detection of the usefulness of heterosis for plant breeding has led to the development of inbred lines that reveal a heterotic yield advantage when they are crossed. Maize was the first species where heterosis was widely used to produce hybrids.

By the 1920s, statistical methods were developed to analyze gene action and distinguish heritable variation from variation caused by environment. In 1933 another important breeding technique, cytoplasmic male sterility (CMS), developed in maize, was described by Marcus Morton Rhoades. CMS is a maternally inherited trait that makes the plant produce sterile pollen. This enables the production of hybrids without the need for labor-intensive detasseling.

These early breeding techniques resulted in large yield increase in the United States in the early 20th century. Similar yield increases were not produced elsewhere until after World War II, the Green Revolution increased crop production in the developing world in the 1960s. This remarkable improvement was based on three essential crops. First came the development of hybrid maize, then high-yielding and input-responsive "semi-dwarf wheat" (for which the CIMMYT breeder N.E. Borlaug received the Nobel prize for peace in 1970), and third came high-yielding "short statured rice" cultivars. Similarly notable improvements were achieved in other crops like sorghum and alfalfa.

Molecular Genetics and Bio-Revolution

Intensive research in molecular genetics has led to the development of recombinant DNA technology (popularly called genetic engineering). Advancement in biotechnological techniques has opened many possibilities for breeding crops. Thus, while mendelian genetics allowed plant breeders to perform genetic transformations in a few crops, molecular genetics has provided the key to both the manipulation of the internal genetic structure, and the "crafting" of new cultivars according to a pre-determined plan.

CLASSICAL PLANT BREEDING

One major technique of plant breeding is selection, the process of selectively propagating plants with desirable characteristics and eliminating or "culling" those with less desirable characteristics.

Another technique is the deliberate interbreeding (crossing) of closely or distantly related individuals to produce new crop varieties or lines with desirable properties. Plants are crossbred to introduce traits/genes from one variety or line into a new genetic background. For example, a mildew-resistant pea may be crossed with a high-yielding but susceptible pea, the goal of the cross being to introduce mildew resistance without losing the high-yield characteristics. Progeny from the cross would then be crossed with the high-yielding parent to ensure that the progeny were most like the high-yielding parent, (backcrossing). The progeny from that cross would then be tested for yield (selection, as described above) and mildew resistance and high-yielding resistant plants would be further developed. Plants may also be crossed with themselves to produce inbred varieties for breeding. Pollinators may be excluded through the use of pollination bags.

Classical breeding relies largely on homologous recombination between chromosomes to generate genetic diversity. The classical plant breeder may also make use of a number of *in vitro* techniques such as protoplast fusion, embryo rescue or mutagenesis (see below) to generate diversity and produce hybrid plants that would not exist in nature.

Traits that breeders have tried to incorporate into crop plants include:

- Improved quality, such as increased nutrition, improved flavor, or greater beauty
- Increased yield of the crop
- Increased tolerance of environmental pressures (salinity, extreme temperature, drought)
- Resistance to viruses, fungi and bacteria
- Increased tolerance to insect pests
- Increased tolerance of herbicides
- Longer storage period for the harvested crop

Before World War II

Successful commercial plant breeding concerns were founded from the late 19th century. Gartons Agricultural Plant Breeders in England was established in the 1890s by John Garton, who was one of the first to commercialize new varieties of agricultural crops created through cross-pollination. The firm's first introduction was Abundance Oat, one of the first agricultural grain varieties bred from a *controlled* cross, introduced to commerce in 1892.

In the early 20th century, plant breeders realized that Mendel's findings on the non-random nature of inheritance could be applied to seedling populations produced through deliberate pollinations to predict the frequencies of different types. Wheat hybrids were bred to increase the crop production of Italy during the so-called "Battle for Grain" (1925–1940). Heterosis was explained by George Harrison Shull. It describes the tendency of the progeny of a specific cross to outperform both parents. The detection of the usefulness of heterosis for plant breeding has led to the development of inbred lines that reveal a heterotic yield advantage when they are crossed. Maize was the first species where heterosis was widely used to produce hybrids.

Statistical methods were also developed to analyze gene action and distinguish heritable variation from variation caused by environment. In 1933 another important breeding technique, cytoplasmic male sterility (CMS), developed in maize, was described by Marcus Morton Rhoades. CMS is a maternally inherited trait that makes the plant produce sterile pollen. This enables the production of hybrids without the need for labor-intensive detasseling.

These early breeding techniques resulted in large yield increase in the United States in the early 20th century. Similar yield increases were not produced elsewhere until after World War II, the Green Revolution increased crop production in the developing world in the 1960s.

After World War II

Following World War II a number of techniques were developed that allowed plant breeders to hybridize distantly related species, and artificially induce genetic diversity.

In vitro-culture of Vitis (grapevine), Geisenheim Grape Breeding Institute.

When distantly related species are crossed, plant breeders make use of a number of plant tissue culture techniques to produce progeny from otherwise fruitless mating. Interspecific and intergeneric hybrids are produced from a cross of related species or genera that do not normally sexually reproduce with each other. These crosses are referred to as *Wide crosses*. For example, the cereal triticale is a wheat and rye hybrid. The cells in the plants derived from the first generation created from the cross contained an uneven number of chromosomes and as result was sterile. The cell division inhibitor colchicine was used to double the number of chromosomes in the cell and thus allow the production of a fertile line.

Failure to produce a hybrid may be due to pre- or post-fertilization incompatibility. If fertilization is possible between two species or genera, the hybrid embryo may abort before maturation. If this does occur the embryo resulting from an interspecific or intergeneric cross can sometimes be rescued and cultured to produce a whole plant. Such a method is referred to as Embryo Rescue. This technique has been used to produce new rice for Africa, an interspecific cross of Asian rice *(Oryza sativa)* and African rice *(Oryza glaberrima)*.

Hybrids may also be produced by a technique called protoplast fusion. In this case protoplasts are fused, usually in an electric field. Viable recombinants can be regenerated in culture.

Chemical mutagens like EMS and DMS, radiation and transposons are used to generate mutants with desirable traits to be bred with other cultivars - a process known as *Mutation Breeding*. Classical plant breeders also generate genetic diversity within a species by exploiting a process called somaclonal variation, which occurs in plants produced from tissue culture, particularly plants derived from callus. Induced polyploidy, and the addition or removal of chromosomes using a technique called chromosome engineering may also be used.

When a desirable trait has been bred into a species, a number of crosses to the favored parent are made to make the new plant as similar to the favored parent as possible. Returning to the example of the mildew resistant pea being crossed with a high-yielding but susceptible pea, to make the mildew resistant progeny of the cross most like the high-yielding parent, the progeny will be crossed back to that parent for several generations. This process removes most of the genetic contribution of the mildew resistant parent. Classical breeding is therefore a cyclical process.

With classical breeding techniques, the breeder does not know exactly what genes have been introduced to the new cultivars. Some scientists therefore argue that plants produced by classical breeding methods should undergo the same safety testing regime as genetically modified plants. There have been instances where plants bred using classical techniques have been unsuitable for human consumption, for example the poison solanine was unintentionally increased to unacceptable levels in certain varieties of potato through plant breeding. New potato varieties are often screened for solanine levels before reaching the marketplace.

MODERN PLANT BREEDING

Modern plant breeding may use techniques of molecular biology to select, or in the case of genetic modification, to insert, desirable traits into plants. Application of biotechnology or molecular biology is also known as molecular breeding (see: Molecular breeding).

Modern facilities in molecular biology have converted classical plant breeding to molecular plant breeding.

Steps of Plant Breeding

The following are the major activities of plant breeding:

- Collection of variation
- Selection
- Evaluation
- Release
- Multiplication
- Distribution of the new variety
- Selling to people

Marker Assisted Selection

Sometimes many different genes can influence a desirable trait in plant breeding. The use of tools such as molecular markers or DNA fingerprinting can map thousands of genes. This allows plant breeders to screen large populations of plants for those that possess the trait of interest. The screening is based on the presence or absence of a certain gene as determined by laboratory procedures, rather than on the visual identification of the expressed trait in the plant.

Reverse Breeding and Doubled Haploids (DH)

A method for efficiently producing homozygous plants from a heterozygous starting plant, which has all desirable traits. This starting plant is induced to produce doubled haploid from haploid cells, and later on creating homozygous/doubled haploid plants from those cells. While in natural offspring genetic recombination occurs and traits can be unlinked from each other, in doubled haploid cells and in the resulting DH plants recombination is no longer an issue. There, a recombination between two corresponding chromosomes does not lead to un-linkage of alleles or traits, since it just leads to recombination with its identical copy. Thus, traits on one chromosome stay linked. Selecting those offspring having the desired set of chromosomes and crossing them will result in a final F1 hybrid plant, having exactly the same set of chromosomes, genes and traits as the starting hybrid plant. The homozygous parental lines can reconstitute the original heterozygous plant by crossing, if desired even in a large quantity. An individual heterozygous plant can be converted into a heterozygous variety (F1 hybrid) without the necessity of vegetative propagation but as the result of the cross of two homozygous/doubled haploid lines derived from the originally selected plant. patent

Genetic Modification

Genetic modification of plants is achieved by adding a specific gene or genes to a plant, or by knocking down a gene with RNAi, to produce a desirable phenotype. The plants resulting from adding a gene are often referred to as transgenic plants. If for genetic modification genes of the species or of a crossable plant are used under control of their native promoter, then they are called cisgenic plants. Sometimes genetic modification can produce a plant with the desired trait or traits faster than classical breeding because the majority of the plant's genome is not altered.

To genetically modify a plant, a genetic construct must be designed so that the gene to be added or removed will be expressed by the plant. To do this, a promoter to drive transcription and a termination sequence to stop transcription of the new gene, and the gene or genes of interest must be introduced to the plant. A marker for the selection of transformed plants is also included. In the laboratory, antibiotic resistance is a commonly used marker: Plants that have been successfully transformed will grow on media containing antibiotics; plants that have not been transformed will die. In some instances markers for selection are removed by backcrossing with the parent plant prior to commercial release.

The construct can be inserted in the plant genome by genetic recombination using the bacteria *Agrobacterium tumefaciens* or *A. rhizogenes*, or by direct methods like the gene gun or microinjection. Using plant viruses to insert genetic constructs into plants is also a possibility, but the technique is limited by the host range of the virus. For example, Cauliflower mosaic virus (CaMV) only infects cauliflower and related species. Another limitation of viral vectors is that the virus is not usually passed on the progeny, so every plant has to be inoculated.

The majority of commercially released transgenic plants are currently limited to plants that have introduced resistance to insect pests and herbicides. Insect resistance is achieved through incorporation of a gene from *Bacillus thuringiensis* (Bt) that encodes a protein that is toxic to some insects. For example, the cotton bollworm, a common cotton pest, feeds on Bt cotton it will ingest the toxin and die. Herbicides usually work by binding to certain plant enzymes and inhibiting their action. The enzymes that the herbicide inhibits are known as the herbicides *target site*. Herbicide resistance can be engineered into crops by expressing a version of *target site* protein that is not inhibited by the herbicide. This is the method used to produce glyphosate resistant crop plants.

Genetic modification of plants that can produce pharmaceuticals (and industrial chemicals), sometimes called *pharming*, is a rather radical new area of plant breeding.

Issues and Concerns

Modern plant breeding, whether classical or through genetic engineering, comes with issues of concern, particularly with regard to food crops. The question of whether breeding can have a negative effect on nutritional value is central in this respect. Although relatively little direct research in this area has been done, there are scientific indications that, by favoring certain aspects of a plant's development, other aspects may be retarded. A study published in the *Journal of the American College of Nutrition* in 2004, entitled *Changes in USDA Food Composition Data for 43 Garden Crops, 1950 to 1999*, compared nutritional analysis of vegetables done in 1950 and in 1999, and found substantial decreases in six of 13 nutrients measured, including 6% of protein and 38% of riboflavin. Reductions in calcium, phosphorus, iron and ascorbic acid were also found. The study, conducted at the Biochemical Institute, University of Texas at Austin, concluded in summary: *"We suggest that any real declines are generally most easily explained by changes in cultivated varieties between 1950 and 1999, in which there may be trade-offs between yield and nutrient content."*

The debate surrounding genetically modified food during the 1990s peaked in 1999 in terms of media coverage and risk perception, and continues today - for example, "*Germany has thrown its weight behind a growing European mutiny over genetically modified crops by banning the planting of a widely grown pest-resistant corn variety.*" The debate encompasses the ecological impact of genetically modified plants, the safety of genetically modified food and concepts used for safety evaluation like substantial equivalence. Such concerns are not new to plant breeding. Most countries have regulatory processes in place to help ensure that new crop varieties entering the marketplace are both safe and meet farmers' needs. Examples include variety registration, seed schemes, regulatory authorizations for GM plants, etc.

Plant breeders' rights is also a major and controversial issue. Today, production of new varieties is dominated by commercial plant breeders, who seek to protect their work and collect royalties through national and international agreements based in intellectual property rights. The range of related issues is complex. In the simplest terms, critics of the increasingly restrictive regulations argue that, through a combination of technical and economic pressures, commercial breeders are reducing biodiversity and significantly constraining individuals (such as farmers) from developing and trading seed on a regional level. Efforts to strengthen breeders' rights, for example, by lengthening periods of variety protection, are ongoing.

When new plant breeds or cultivars are bred, they must be maintained and propagated. Some plants are propagated by asexual means while others are propagated by seeds. Seed propagated cultivars require specific control over seed source and production procedures to maintain the integrity of the plant breeds results. Isolation is necessary to prevent cross contamination with related plants or the mixing of seeds after harvesting. Isolation is normally accomplished by planting distance but in certain crops, plants are enclosed in greenhouses or cages (most commonly used when producing F1 hybrids.)

ROLE OF PLANT BREEDING IN ORGANIC AGRICULTURE

Critics of organic agriculture claim it is too low-yielding to be a viable alternative to conventional agriculture. However, part of that poor performance may be the result of growing poorly adapted varieties. It is estimated that over 95% of organic agriculture is based on conventionally adapted varieties, even though the production environments found in organic vs. conventional farming systems are vastly different due to their distinctive management practices. Most notably, organic farmers have fewer inputs available than conventional growers to control their production environments. Breeding varieties specifically adapted to the unique conditions of organic agriculture is critical for this sector to realize its full potential. This requires selection for traits such as:

- Water use efficiency
- Nutrient use efficiency (particularly nitrogen and phosphorus)
- Weed competitiveness
- Tolerance of mechanical weed control

- Pest/disease resistance
- Early maturity (as a mechanism for avoidance of particular stresses)
- Abiotic stress tolerance (i.e. drought, salinity, etc...)

Currently, few breeding programs are directed at organic agriculture and until recently those that did address this sector have generally relied on indirect selection (i.e. selection in conventional environments for traits considered important for organic agriculture). However, because the difference between organic and conventional environments is large, a given genotype may perform very differently in each environment due to an interaction between genes and the environment. If this interaction is severe enough, an important trait required for the organic environment may not be revealed in the conventional environment, which can result in the selection of poorly adapted individuals. To ensure the most adapted varieties are identified, advocates of organic breeding now promote the use of direct selection (i.e. selection in the target environment) for many agronomic traits.

There are many classical and modern breeding techniques that can be utilized for crop improvement in organic agriculture despite the ban on genetically modified organisms. For instance, controlled crosses between individuals allow desirable genetic variation to be recombined and transferred to seed progeny via natural processes. Marker assisted selection can also be employed as a diagnostics tool to facilitate selection of progeny who possess the desired trait(s), greatly speeding up the breeding process. This technique has proven particularly useful for the introgression of resistance genes into new backgrounds, as well as the efficient selection of many resistance genes pyramided into a single individual. Unfortunately, molecular markers are not currently available for many important traits, especially complex ones controlled by many genes.

Addressing Global Food Security Through Plant Breeding

For future agriculture to thrive there are necessary changes which must be made in accordance to arising global issues. These issues are arable land, harsh cropping conditions and food security which involves, being able to provide the world population with food containing sufficient nutrients. These crops need to be able to mature in several environments allowing for worldwide access, this is involves issues such as drought tolerance. These global issues are achievable through the process of plant breeding, as it offers the ability to select specific genes allowing the crop to perform at a level which yields the desired results.

Minimal Land Degradation

Land degradation is a major issue, as it can negatively impact the capability of the land to be productive. Poor agricultural management has a huge impact on the degradation of soil worldwide and it is Africa and Asia that are most affected. Through education and development of modified plants, these statistics can be reduced and agricultural land can become more productive. Plant breeding allows for an increase in yield with out the extra strain on the land. The genetically modified, Bt white maize, was introduced to South Africa and was surveyed in 33 large commercial farms and 368 small landholders properties and in both cases a higher yield was recorded.

Increased Yield without Expansion

With an increasing population, the production of food needs to increase with it. It is estimated that a 70% increase in food production is needed by 2050 in order to meet the Declaration of the World Summit on Food Security. But with the natural degradation of agricultural land, simply planting more crops is no longer a viable option. Therefore, new varieties of plants need to be developed through plant breeding that generates an increase of yield without relying on an increase in land area. An example of this can be seen in Asia, where food production per capita has increased twofold. This has been achieved through not only the use of fertilizers, but through the use of better crops that have been specifically designed for the area.

Breeding for Increased Nutritional Value

Plant breeding can contribute to global food security as it is a cost-effective tool for increasing nutritional value of forage and crops. Improvements in nutritional value for forage crops from the use of analytical chemistry and rumen fermentation technology have been recorded since 1960; this science and technology gave breeders the ability to screen thousands of samples within a small amount of time, meaning breeders could identify a high performing hybrid quicker. The main area genetic increases were made was in vitro dry matter digestibility (IVDMD) resulting in 0.7-2.5% increase, at just 1% increase in IVDMD a single Bos Taurus also known as beef cattle reported 3.2% increase in daily gains. This improvement indicates plant breeding is an essential tool in gearing future agriculture to perform at a more advanced level.

Breeding for Tolerance

Plant breeding of hybrid crops has become extremely popular worldwide in an effort to combat the harsh environment. With long periods of drought and lack of water or nitrogen stress tolerance has become a significant part of agriculture. Plant breeders have focused on identifying crops which will ensure crops perform under these conditions; a way to achieve this is finding strains of the crop that is resistance to drought conditions with low nitrogen. It is evident from this that plant breeding is vital for future agriculture to survive as it enables farmers to produce stress resistant crops hence improving food security.

Participatory Plant Breeding

Participatory plant breeding (PPB) is when farmers are involved in a crop improvement programme with opportunities to make decisions and contribute to the research process at different stages. Participatory approaches to crop improvement can also be applied when plant biotechnologies are being used for crop improvement.

REFERENCES

1. Breeding Field Crops. 1995. Sleper and Poehlman. Page 3

2. Piperno, D. R.; Ranere, A. J.; Holst, I.; Iriarte, J.; Dickau, R. (2009). "Starch grain and phytolith evidence for early ninth millennium B.P. maize from the Central Balsas River Valley, Mexico". PNAS. 106 (13): 5019–5024. doi:10.1073/pnas.0812525106. PMC 2664021Freely accessible. PMID 19307570.

3. Deppe, Carol (2000). Breed Your Own Vegetable Varieties. Chelsea Green Publishing. |page=237-244

4. "Plant breeding".

5. Spring Seed Catalogue 1899, Gartons Limited

6. Noel Kingsbury (2009). Hybrid: The History and Science of Plant Breeding. University of Chicago Press. p. 140.

7. Suzie Key; Julian K-C Ma & Pascal MW Drake (1 June 2008). "Genetically modified plants and human health". Journal of the Royal Society of Medecine. 101: 290–298. doi:10.1258/jrsm.2008.070372. PMC 2408621Freely accessible. PMID 18515776.

8. Davis, D.R.; Epp, M.D.; Riordan, H.D. (2004). "Changes in USDA Food Composition Data for 43 Garden Crops, 1950 to 1999". Journal of the American College of Nutrition. 23 (6): 669–682. doi:10.1080/07315724.2004.10719409.

9. Costa-Font, J.; Mossialos, E. (2007). "Are perceptions of 'risks' and 'benefits' of genetically modified food (in)dependent?". Food Quality and Preference. 18: 173–182. doi:10.1016/j.foodqual.2005.09.013.

10. Connoly, Kate (2009-04-14). "Germany deals blow to GM crops". The Guardian. Retrieved 2009-06-25.

11. Murphy, Kevin M.; K.G. Campbell; S.R. Lyon; S.S. Jones (2007). "Evidence of varietal adaptation to organic farming systems". Field Crops Research. 102 (3): 172–177. doi:10.1016/j.fcr.2007.03.011.

12. Lammerts van Bueren, E.T.; S.S. Jones; L. Tamm; K.M. Murphy; J.R. Myers; C. Leifert; M.M. Messmer (2010). "The need to breed crop varieties suitable for organic farming, using wheat, tomato and broccoli as examples: A review". NJAS- Wageningen Journal of Life Sciences. 58 (3–4): 193–205. doi:10.1016/j.njas.2010.04.001.

13. Lammerts van Bueren, E. T.; G. Backes; H. de Vriend; H. Ostergard (2010). "The role of molecular markers and marker assisted selection in breeding for organic agriculture". Euphytica. 175: 51–64. doi:10.1007/s10681-010-0169-0.

14. Oldeman, l (1994). "The global extent of soil degradation" (PDF). Soil resilience and sustainable land use. 32 (5967): 818–822. Retrieved 7-11-2013. Check date values in: |access-date= (help)

15. Tester, Mark; Langridge, Peter (February 2010). "Breeding technologies to increase crop production in a changing world". Science. AAAS: American Association for the Advancement of Science. 327 (5967): 818–822. doi:10.1126/science.1183700.

16. Haddad, Lawrence; Godfray, H. Charles J.; Beddington, John R.; Crute, Ian R.; Lawrence, David; Muir, James F.; Pretty, Jules; Robinson, Sherman; Thomas,

Sandy M. and Toulmin, Camilla (12 February 2010). "Food security: the challenge of feeding 9 billion people". Science. AAAS: American Association for the Advancement of Science. 327 (5967): 812–818. doi:10.1126/science.1185383. PMID 20110467. Cite uses deprecated parameter |coauthor= (help)

17. Bänziger (2000). "Breeding for drought and nitrogen stress tolerance in maize: from theory to practice". From Theory to Practice: 7–9. Retrieved 7-11-2013. Check date values in: |access-date= (help)
18. Casler, Vogal, M,K (1998-03-31). "Accomplishments and impact from breeding for increased forage nutritional value". Crop Science. 39 (1): 12–20. doi:10.2135/cropsci1999.0011183x003900010003x. Retrieved 7-10. Check date values in: |access-date=, |year= / |date= mismatch (help)
19. PRGA 2008. Participatory Plant Breeding (PPB)
20. Sperling et al. 2001. A Framework For Analizing Participatory Plant Breeding Approaches And Results.
21. Ceccarelli 2001. Decentralized-Participatory Plant Breeding: Adapting Crops to Environments and Clients.
22. Thro A & Spillane C (2000). "Biotechnology-assisted Participatory Plant Breeding: Complement or Contradiction?" (PDF). CGIAR Systemwide Program on Participatory Research and Gender Analysis for Technology Development and Institutional Innovation. Working Document No. 4 April 2000: 140.
23. Borem, A.; Miranda, G. V. Melhoramento de Plantas. 5ª. ed. Viçosa: Editora UFV, 2009. v. 1. 543 p.
24. Borem, A. (Org.). Domesticação e Melhoramento: espécies amazônicas1. ed. Visconde do Rio Branco: Suprema Grafica e Editora, 2009. (in press)v. 1. 588 p.
25. Borem, A. (Org.); Caixeta, E. T. (Org.) . Marcadores Moleculares. 2a.. ed. Visconde do Rio Branco: Suprema Grafica e Editora, 2008. v. 1. 532 p.
26. Borem, A.; Condori, M.; Miranda, G. V. Mejoramiento de Plantas (in Spanish). 1. ed. Viçosa: Editora UFV, 2008. v. 1. 438 p.
27. McCouch, S. (2004). "Diversifying Selection in Plant Breeding". PLoS Biol. 2 (10): e347. doi:10.1371/journal.pbio.0020347. PMC 521731Freely accessible. PMID 15486582.
28. Briggs, F.N. and Knowles, P.F. 1967. Introduction to Plant Breeding. Reinhold Publishing Corporation, New York.
29. Gepts, P. (2002). "A Comparison between Crop Domestication, Classical Plant Breeding, and Genetic Engineering". Crop Science. 42 (6): 1780–1790. doi:10.2135/cropsci2002.1780.
30. The Origins of Agriculture and Crop Domestication - The Harlan Symposium
31. news@nature.com. 1999 Are non-GM crops safe?
32. Schlegel,Rolf (2009) Encyclopedic Dictionary of Plant Breeding 2nd ed. (ISBN 9781439802427), CRC Press, Boca Raton, FL, USA, pp 584

33. Schlegel, Rolf (2007) Concise Encyclopedia of Crop Improvement: Institutions, Persons, Theories, Methods, and Histories (ISBN 9781560221463), CRC Press, Boca Raton, FL, USA, pp 423
34. Schouten, Henk J.; Krens, Frans A.; Jacobsen, Evert (2006). "Do cisgenic plants warrant less stringent oversight?". Nature Biotechnology. 24 (7): 753. doi:10.1038/nbt0706-753. PMID 16841052.
35. Schouten, Henk J.; Krens, Frans A; Jacobsen, Evert (2006). "Cisgenic plants are similar to traditionally bred plants". EMBO Reports. 7 (8): 750–753. doi:10.1038/sj.embor.7400769. PMC 1525145Freely accessible. PMID 16880817.
36. Sun. "From indica and japonica splitting in common wild rice DNA to the origin and evolution of Asian cultivated rice". Agricultural Archaeology. 1998: 21–29.
37. Thro AM and Spillane C (1999) Biotechnology assisted participatory plant breeding: Complement or contradiction? CGIAR Program on Participatory Research and Gender Analysis, Working Document No.4, CIAT: Cali. 150pp.
38. Deppe, Carol (2000). Breed Your Own Vegetable Varieties. Chelsea Green Publishing.

CHAPTER 2

Plant Cultigen and Cultivars

Chapter Contents

- **Introduction**
- **Origin of Term**
- **Different Kinds**

INTRODUCTION

The term *cultivar* is, most commonly, referring to an assemblage of plants selected for desirable characteristics that are maintained during propagation. More generally, *cultivar* refers to the most basic classification category of cultivated plants governed by the ICNCP. Most cultivars have arisen in cultivation, but a few are special selections from the wild.

Popular ornamental garden plants like roses, camellias, daffodils, rhododendrons, and azaleas are cultivars produced by careful breeding and selection for flower color and form. Similarly, the world's agricultural food crops are almost exclusively cultivars that have been selected for characteristics such as improved yield, flavor, and resistance to disease and very few wild plants are now used as food sources. Trees used in forestry are also special selections grown for their enhanced quality and yield of timber.

Cultivars form a major part of Liberty Hyde Bailey's broader grouping, the cultigen, defined as a plant whose origin or selection is primarily due to intentional human activity. *Cultivar* was coined by Bailey and it is generally regarded as a portmanteau of "cultivated" and "variety", but could also be derived from "cultigen" and "variety". A cultivar is not the same as a botanical variety, a taxonomic rank below subspecies, and there are differences in the rules for creating and using the names of botanical varieties and cultivars. In recent times, the naming of cultivars has been complicated by the use of statutory plant patents and plant breeders' rights names.

The International Union for the Protection of New Varieties of Plants (UPOV – French: *Union internationale pour la protection des obtentions végétales*) offers legal protection of plant cultivars to people or organizations who introduce new cultivars to commerce. UPOV requires that a cultivar be *distinct, uniform* and *stable*. To be distinct, it must have characteristics that easily distinguish it from any other known cultivar. To be uniform and stable, the cultivar must retain these characteristics under repeated propagation.

Osteospermum 'Pink Whirls' A cultivar selected for its intriguing and colorful flowers.

The naming of cultivars is an important aspect of cultivated plant taxonomy, and the correct naming of a cultivar is prescribed by the Rules and Recommendations of the *International Code of Nomenclature for Cultivated Plants* (the *ICNCP*, commonly known as the *Cultivated Plant Code*). A cultivar is given a cultivar name, which consists of the scientific Latin botanical name followed by a cultivar epithet. The cultivar epithet is usually in a vernacular language. For example, the full cultivar name of the King Edward potato is *Solanum tuberosum* 'King Edward'. The 'King Edward' part of the name is the cultivar epithet, which, according to the Rules of the *Cultivated Plant Code*, is bounded by single quotation marks.

ORIGIN OF TERM

The origin of the term "cultivar" arises from the need to distinguish between wild plants and those with characteristics that have arisen in cultivation (what we now call cultigens). This distinction dates back to the Greek philosopher Theophrastus (370–285 BCE), the

"Father of Botany", who was keenly aware of this difference. Botanical historian Alan Morton notes that Theophrastus in his *Enquiry into Plants* "*had an inkling of the limits of culturally induced (phenotypic) changes and of the importance of genetic constitution*" (*Historia Plantarum* III, 2,2 and *Causa Plantarum* I, 9,3).

Liberty Hyde Bailey (1858–1954) coined the word *cultigen* in 1918 and *cultivar* in 1923.

The International Code of Nomenclature for algae, fungi, and plants uses as its starting point for modern botanical nomenclature those Latin names that appeared in Linnaeus' publications *Species Plantarum* (10th ed.) and *Genera Plantarum* (5th ed.). In *Species Plantarum*, Linnaeus (1707–1778) listed all the plants known to him, either directly or from his extensive reading. He recognised the rank of varietas (in English this is the botanical "variety", a rank below that of species and subspecies) and he indicated these varieties by using letters of the Greek alphabet such as α, β, λ in front of the variety name, rather than using the abbreviation var., which is the current convention. Most of the varieties listed by Linnaeus were of "garden" origin rather than being wild plants.

Over time there was an increasing need to distinguish between plants growing in the wild, and those with variations that had been produced in cultivation. In the nineteenth century many "garden-derived" plants were given horticultural names, sometimes in Latin and sometimes in a local language. From about the 1900s, plants produced in cultivation in Europe were recognized in the Scandinavian, Germanic, and Slavic literature through the words *stamm* or *sorte* but these words could not be used internationally since, by international agreement, any new terms had to be based in Latin. In the twentieth century an improved international terminology was proposed for the classification and nomenclature of cultivated plants.

The word *cultivar* was coined in 1923 by Liberty Hyde Bailey of Cornell University, New York State, when he wrote:

The cultigen is a species, or its equivalent, that has appeared under domestication – the plant is cultigenous. I now propose another name, cultivar, for a botanical variety, or for a race subordinate to species, that has originated under cultivation; it is not necessarily, however, referable to a recognized botanical species. It is essentially the equivalent of the botanical variety except in respect to its origin.

In this paper Bailey used only the rank of species for the cultigen but it was clear to him that many domesticated plants were more like botanical varieties than species, and that appears to have motivated the suggestion of the new classification category *cultivar*, which is generally assumed to be a contraction of the words *cultivated* and *variety*. However, Bailey was never explicit about the etymology of the word, and it has been suggested that it is a contraction of the words *cultigen* and *variety*, which seems more appropriate.

The new word *cultivar* was promoted as "euphonious" and "free from ambiguity". It serves a purpose. Its use was subsequently recommended by the first *Cultivated Plant Code*, which was published in 1953, and by 1960 it had achieved wide international acceptance.

Cultigens

The terms *cultigen* and *cultivar* may be confused with each other. A *cultigen* is any plant deliberately altered or selected for cultivation by humans; the *Cultivated Plant Code* states that cultigens are "maintained as recognizable entities solely by continued propagation". A cultigen or a component of a cultigen, which is recognizable and stable, can be accepted as a *cultivar.*

Bread wheat, *Triticum aestivum*, is considered a cultigen, and is a distinct species from other wheats according to the biological species concept. Many separate cultivars have been created within the cultigen. Many other cultigens are not considered to be distinct species, and can be named in other ways.

Cultigens can have names at any of various ranks, including cultivar names, names in the classification categories of grex and group, binomial Latin species names, variety names, forma names, and they may be plants that have been altered by humans (including genetically modified plants) but which have not been given formal names.

Formal Definition

The *Cultivated Plant Code* notes that the word cultivar is used in two different senses: first, as a "classification category" the cultivar is defined in Article 2 of the *International Code of Nomenclature for Cultivated Plants* (2009, 8th edition) as follows: *The basic category of cultivated plants whose nomenclature is governed by this Code is the cultivar.* There are two other classification categories for cultigens, the grex and the group. The *Code* then defines a *cultivar* as a "taxonomic unit within the classification category of cultivar". This is the sense of *cultivar* that is most generally understood and which is used as a general definition.

A cultivar is an assemblage of plants that (a) has been selected for a particular character or combination of characters, (b) is distinct, uniform and stable in those characters, and (c) when propagated by appropriate means, retains those characters.

DIFFERENT KINDS

Which plants are chosen to be named as cultivars is simply a matter of convenience as the category was created to serve the practical needs of horticulture, agriculture, and forestry.

A cultivar of the orchid genus *Oncidium*.

Members of a particular cultivar are not necessarily genetically identical. The *Cultivated Plant Code* emphasizes that different cultivated plants may be accepted as different cultivars, even if they have the same genome, while cultivated plants with different genomes may be regarded as the same cultivar. The production of cultivars generally entails considerable human involvement although in a few cases it may be as little as simply selecting variation from plants growing in the wild (whether by collecting growing tissue to propagate from or by gathering seed).

Cultivars generally occur as ornamentals and food crops: *Malus* 'Granny Smith' and *Malus* 'Red Delicious' are cultivars of apples propagated by cuttings or grafting, *Lactuca*

'Red Sails' and *Lactuca* 'Great Lakes' are lettuce cultivars propagated by seeds. Named cultivars of *Hosta* and *Hemerocallis* plants are cultivars produced by micropropagation or division.

Clones

Leucospermum 'Scarlet Ribbon'

A cross performed in Tasmania between *L. glabrum* and *L. tottum*

Cultivars that are produced asexually are genetically identical and known as clones; this includes plants propagated by division, layering, cuttings, grafts, and budding. The propagating material may be taken from a particular part of the plant, such as a lateral branch, or from a particular phase of the life cycle, such as a juvenile leaf, or from aberrant growth as occurs with witch's broom. Plants whose distinctive characters are derived from the presence of an intracellular organism may also form a cultivar provided the characters are reproduced reliably from generation to generation. Plants of the same chimera (which have mutant tissues close to normal tissue) or graft-chimeras (which have vegetative tissue from different kinds of plants and which originate by grafting) may also constitute a cultivar.

Seed-Produced

Some cultivars "come true from seed", retaining their distinguishing characteristics when grown from seed. Such plants are termed a "variety", "selection" or "strain" but these are ambiguous and confusing words that are best avoided. In general, asexually propagated cultivars grown from seeds produce highly variable seedling plants, and should not be labelled with, or sold under, the parent cultivar's name.

Seed-raised cultivars may be produced by uncontrolled pollination when characteristics that are distinct, uniform and stable are passed from parents to progeny. Some are produced as "lines" that are produced by repeated self-fertilization or inbreeding or "multilines" that are made up of several closely related lines. Sometimes they are F1 hybrids which are the result of a deliberate repeatable single cross between two pure lines. A few F2 hybrid seed cultivars also exist, such as *Achillea* 'Summer Berries'.

Some cultivars are agamospermous plants, which retain their genetic composition and characteristics under reproduction. Occasionally cultivars are raised from seed of a specially selected provenance – for example the seed may be taken from plants that are resistant to a particular disease.

Genetically Modified

Genetically modified plants with characteristics resulting from the deliberate implantation of genetic material from a different germplasm may form a cultivar. However, the International Code of Nomenclature for Cultivated Plants notes, "In practice such an assemblage is often marketed from one or more lines or multilines that have been genetically modified. These lines or multilines often remain in a constant state of development which makes the naming of such an assemblage as a cultivar a futile exercise." However, retired transgenic varieties such as the Fish tomato, which are no longer being developed, do not run into this obstacle and can be given a cultivar name.

Cultivars may be selected because of a change in the ploidy level of a plant which may produce more desirable characteristics.

Cultivar Names

Viola 'Clear Crystals Apricot'

The specific epithet may be omitted from a cultivar name

Every unique cultivar has a unique name within its denomination class (which is almost always the genus). Names of cultivars are regulated by the *International Code of Nomenclature for Cultivated Plants*, and may be registered with an International Cultivar Registration Authority (ICRA). There are sometimes separate registration authorities for different plant types such as roses and camellias. In addition, cultivars may be associated with commercial marketing names referred to in the *Cultivated Plant Code* as "trade designations" (see below).

Presenting in Text

A *cultivar name* consists of a botanical name (of a genus, species, infraspecific taxon, interspecific hybrid or intergeneric hybrid) followed by a cultivar epithet. The cultivar epithet is enclosed by single quotes; it should not be italicized if the botanical name is italicized; and each of the words within the epithet is capitalized (with some permitted exceptions such as conjunctions). It is permissible to place a cultivar epithet after a common name provided the common name is botanically unambiguous. Cultivar epithets published before 1 January 1959 were often given a Latin form and can be readily confused with the specific epithets in botanical names; after that date, newly coined cultivar epithets must be in a modern vernacular language to distinguish them from botanical epithets.

Examples of correct text presentation:

- *Cryptomeria japonica* 'Elegans'
- *Chamaecyparis lawsoniana* 'Aureomarginata' (pre-1959 name, Latin in form)
- *Chamaecyparis lawsoniana* 'Golden Wonder' (post-1959 name, English language)
- *Pinus densiflora* 'Akebono' (post-1959 name, Japanese language)

Apple 'Sundown'

Some incorrect text presentation examples:

- *Cryptomeria japonica* "Elegans" (double quotes are unacceptable)
- *Berberis thunbergii* cv. 'Crimson Pygmy' (this once-common usage is now unacceptable, as it is no longer correct to use "cv." in this context; *Berberis thunbergii* 'Crimson Pygmy' is correct)
- *Rosa* cv. 'Peace' (this is now incorrect for two reasons: firstly, the use of "cv."; secondly, "Peace" is a trade designation or "selling name" for the cultivar *R.* 'Madame A. Meilland' and should therefore be printed in a different typeface from the rest of the name, without quote marks, for example: *Rosa* Peace.)

Group Names

Where several very similar cultivars exist they can be associated into a *Group* (formerly *Cultivar-group*). As Group names are used with cultivar names it is necessary to understand their way of presentation. Group names are presented in normal type and the first letter of each word capitalised as for cultivars, but they are not placed in single quotes. When used in a name, the first letter of the word "Group" is itself capitalized.

Presenting in Text

- *Brassica oleracea* Capitata Group (the group of cultivars including all typical cabbages)
- *Brassica oleracea* Botrytis Group (the group of cultivars including all typical cauliflowers)
- *Hydrangea macrophylla* Groupe Hortensia (in French) = *Hydrangea macrophylla* Hortensia Group (in English)

Where cited with a cultivar name the group should be enclosed in parentheses, as follows:

- *Hydrangea macrophylla* (Hortensia Group) 'Ayesha'

Legal Protection of Cultivars and their Names

Intellectual property law and Intellectual rights
Primary rights
• Copyright • Patent • Trademark • Trade dress • Trade secret • Geographical indication
***Sui generis* rights**
• Authors' rights • Database right • Indigenous intellectual property • Industrial design right • Integrated circuit layout design protection • Moral rights • Plant breeders' rights • Related rights • Supplementary protection certificate • Utility model
Related topics

- Abandonware
- Anti-copyright
- Bioprospecting
- Limitations and exceptions to copyright
- (Fair dealing
- Fair use
- Right to quote)
- Orphan works
- Public domain
- Societal views
- Traditional safety valves
- Outline of intellectual property

Since the 1990s there has been an increasing use of legal protection for newly produced cultivars. Plant breeders expect legal protection for the cultivars they produce. If other growers can immediately propagate and sell these cultivars as soon as they come on the market, the breeder's benefit is largely lost. Legal protection for cultivars is obtained through the use of Plant breeders' rights and plant Patents but the specific legislation and procedures needed to take advantage of this protection vary from country to country.

Controversial Use of Legal Protection for Cultivars

The use of legal protection for cultivars can be controversial, particularly for food crops that are staples in developing countries, or for plants selected from the wild and propagated for sale without any additional breeding work; some people consider this practice unethical.

Trade Designations

The formal scientific name of a cultivar, like *Solanum tuberosum* 'King Edward', is a way of uniquely designating a particular kind of plant. This scientific name is in the public domain and cannot be legally protected. Plant retailers wish to maximize their share of the market and one way of doing this is to replace the cumbersome Latin scientific names on plant labels in retail outlets with appealing marketing names that are easy to use, pronounce and remember. Marketing names lie outside the scope of the *Cultivated Plant Code* which refers to them as "trade designations". If a retailer or wholesaler has the sole legal rights to a marketing name then that may offer a sales advantage. Plants protected by Plant breeders' rights (PBR) may have a "true" cultivar name – the recognized scientific name in the public domain, and a "commercial synonym" an additional marketing name that is legally protected: an example would be *Rosa* Fascination = 'Poulmax', the 'Poulmax' being the true scientific name. Because a name that is attractive in one language may have less appeal in another country, a plant may be given different selling names from country to country. Quoting the original cultivar name allows the correct identification of cultivars around the world. The peak body coordinating Plant breeders rights is the Union for the Protection of New Varieties of Plants (*Union internationale pour la protection des obtentions végétales*, UPOV) and this organization maintains a database of new cultivars protected by PBR in all countries.

International Cultivar Registration Authorities

An *International Cultivar Registration Authority* (ICRA) is a voluntary, non-statutory organization appointed by the *Commission for Nomenclature and Cultivar Registration* of the International Society of Horticultural Science. ICRAs are generally formed by societies and institutions specializing in particular plant genera such as *Dahlia* or *Rhododendron* and are currently located in Europe, North America, China, India, Singapore, Australia, New Zealand, South Africa and Puerto Rico.

Dahlia 'Akita' A cultivar selected for flower form and color.

Each ICRA produces an annual report and its reappointment is considered every four years. The main task is to maintain a register of the names within the group of interest and where possible this is published and placed in the public domain. One major aim is to prevent the duplication of cultivar and Group epithets within a genus, as well as ensuring that names are in accord with the latest edition of the *Cultivated Plant Code*. In this way, over the last 50 years or so, ICRAs have contributed to the stability of cultivated plant nomenclature. In recent times many ICRAs have also recorded trade designations and trademarks used in labelling plant material, to avoid confusion with established names.

New names and other relevant data are collected by and submitted to the ICRA and in most cases there is no cost. The ICRA then checks each new epithet to ensure that it has not been used before and that it conforms with the *Cultivated Plant Code*. Each ICRA also ensures that new names are formally established (i.e. published in hard copy, with a description in a dated publication). They record details about the plant, such as parentage, the names of those concerned with its development and introduction, and a basic description highlighting its distinctive characters. ICRAs are not responsible for assessing the distinctiveness of the plant in question. Most ICRAs can be contacted electronically and many maintain web sites: for an up-to-date listing.

List of Mango Cultivars

Worldwide, hundreds of mango cultivars exist. In mango orchards, multiple cultivars are often grown together to improve cross-pollination.

Table of Mangoes

Reportedly, in India alone, there are around 283 types of mangoes, out of which only 30 are well-known. The following are among the more widely grown mango cultivars, listed by the country in which they were selected or are most extensively cultivated:

Common name(s)	Image	Origin/ region	Notes
Alampur Baneshan		India, United States	
Alice		United States	
Almaas		Pakistan	
Alphonso		India, Pakistan, Ecuador, Egypt, Sudan, United States, Nepal	"King of Mangoes", very sweet with fibreless pulp, rich in vitamin A and C, founded mainly in Devgad & Ratnagiri Districts of Maharashtra, India. This cultivar is available in Feb till end of May.
Amrapali		India	

Anderson	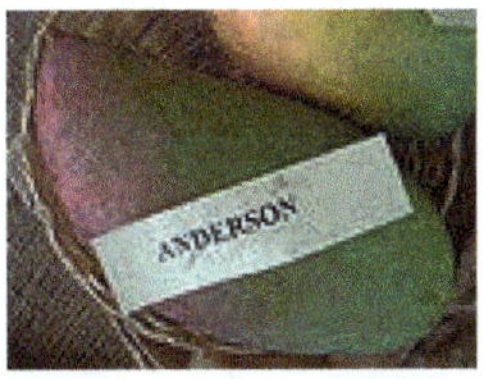	United States	
Angie		United States	
Anwar Ratol		Multan, Punjab (Pakistan)	Small in size and flesh without fibre . Peak season is mid July to August.
Ataulfo		Ecuador, Mexico	
Bailey's Marvel		United States	
Banganapalli	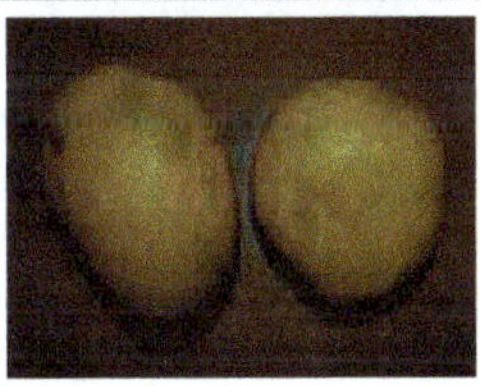	India	The largest volume mango cultivar in the world, this medium-large fruit ripens to a golden-yellow exterior and a straw-yellow to golden-yellow interior. Very juicy, slightly tangy with little or no fibre and a creamy texture. It originates from the erstwhile princely state of Banganapalle, in present-day Andhra Pradesh, India.
Bennet Alphonso		United States	

Beverly		United States	
Black and Rose		India Kerala	Black and Rose
Bombay		India, Nepal, United States	
Brahm Kai Meu		United States	
Brooks		Australia, United States	
Carabao (Philippine Mango)		Philippines	The Ataulfo (mango) and Manilita mango cultivars originated from the Philippine Mango variety. It can be traced back in the Manila-Mexico galleon trade in the years 1600-1800's.
Carrie		United States	
Chaunsa		India, Sindh, Pakistan	

Chok anan		Pakistan, Bangladesh, India, Thailand	
Cogshall		United States	
Cushman		United States	
Dasheri		India, Nepal, Pakistan,	stories about its origin but the owners say it is a "chance" seedling.
Dot		United States	
Dudhiya Malda		India, Nepal	Grown in Digha, Patna, Bihar. Known for awesome taste and flavour with thin skin
Duncan		United States	
Earlygold		United States	
Edward		United States	

Eldon		United States	
Emerald		United States	
Fazli (mango)		Bangladesh, India	
Fajri Kalan		Pakistan	
Fairchild		United States	
Fascell		United States	
Florigon		United States	
Ford		United States	
Gary		United States	
Gir Kesar		Gujarat (India)	
Glenn		Italy, United States	
Gold Nugget		United States	

Golden Lippens		United States	
Graham		United States	
Green Willard		Sri Lanka	
Haden	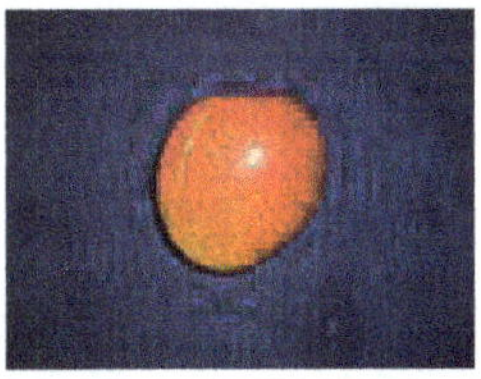	Australia, Brazil, Costa Rica, Ecuador, Guatemala, Honduras, Mexico, United States	
Hatcher		United States	
Heidi		South Africa	
Himayat / Imam Pasand		India	
Himsagar		Bangladesh, India, Nepal	
Ice Cream		United States	
Irwin		Australia, Costa Rica, United States, Taiwan, Japan (Okinawa)	In Japan this variety is marketed as Apple Mango, and in the case of imports from Japan one needs to take care not to confuse it with the Apple Mango grown in the Philippines.
Ivory		China	Also known as the Jingu Ivory mango, or Ivory mango, this long, thin mango is named for its resemblance to a young elephant's tusk. It has thin skin, smooth skin. The flesh contains very few fibres, and constitutes approximately 82 percent of the fruit. It was first introduced into Yunnan, China from Thailand in 1914. The actual tree that was the first to be imported still grows, and during one year produced almost 500 kg of fruit.

Jakarta		United States	
Jean Ellen		United States	
Julie		Ecuador, United States	A colourful, potato shaped mango with a sweet-tart flavour and a tall growth habit. Very popular in the Caribbean Sea, including Jamaica the British Virgin Islands, the United States Virgin Islands, and Barbados.
Kalepad		Andhra Pradesh, India	
Keitt		Australia, Costa Rica, Ecuador, Italy, South Africa, United States	
Kensington Pride		Australia, Italy, United States	
Kent		Australia, Ecuador, Guatemala, Honduras, Israel, Italy, Mexico, South Africa, United States	
Kohu Amba		Sri Lanka	
Lakshmanbhog		India	
Lancetilla		Honduras, United States	

Langra		India, Pakistan, Bangladesh,	
Lippens		United States	
Madame Francis		Haiti	Large kidney shaped mango that ripens to a golden yellow colour with piquant flesh Commonly exported to the United States in spring; often a feature of NYC fruit stands.
Mallika		India, Nepal, United States	
Malwana (mango)		Sri Lanka	
Manilita		United States	
Manohar		Punjab, India	Large mango grows in clusters, seedling of Chaunsa. Exceptional complex flavor good sweetness. Very vigorous.
Momi K		Hawaii, United States	Teardrop shaped with a narrow point. Ripens to a reddish orange blush. Flesh has creamy consistency.
Muhammad Wala		Pakistan	
Mulgoba		India, United States	
Neelam		India, Pakistan	

Osteen		Italy, Spain, United States	
Palmer		Australia, Brazil, United States	
Parvin		United States	
Raspuri		Karnataka, India	An extremely popular variety in South India, a fully ripe Raspuri mango harvested at the right time and ripened naturally can beat them all, including the Alphonso, in taste as well as amount of juice per mango. Many people especially Kannadigas do not even ask for any another variety except Raspuri in Bengaluru.
Red Willard		Sri Lanka	
Rosigold		United States	
Ruby		United States	

Saigon		United States	
Sindhri	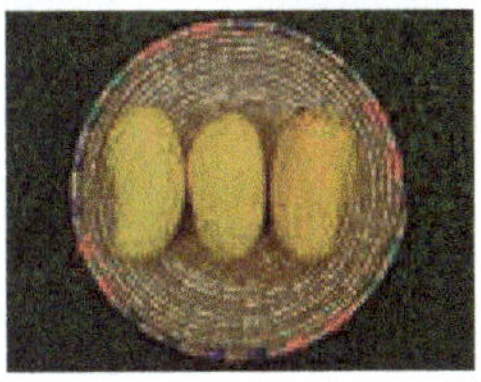	Sindh Pakistan	Sindhi Mango cultivated in Sindhri town Mirpur Khas district, Sindh province of Pakistan
Saharni		Pakistan	
Sammar Bahisht		Pakistan, India	(in Urdu Sammar means fruit and Bahisht means Paradise)
Sensation		South Africa, United States	
Shan-e-Khuda		Pakistan	
Sindhri	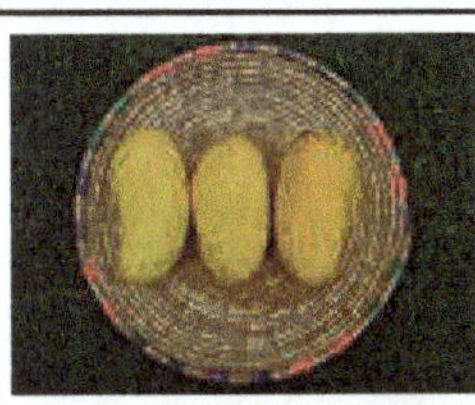	Mirpur Khas district, Sindh province of Pakistan	Sindhi Mango cultivated in Sindhri town Mirpur Khas district, Sindh province of Pakistan
Sophie Fry		United States	
Southern Blush		United States	
Spirit of '76		United States	

Springfels		United States	
Sunset		United States	
Tommy Atkins		Brazil, Costa Rica, Ecuador, Guatemala, Honduras, Israel, Italy, Mexico, South Africa, United States, Venezuela	One of the most commonly sold mangoes in the world. has extremely long shelf life. Hard and oblong exterior.
Torbert		United States	
Totapuri		India	
Valencia Pride	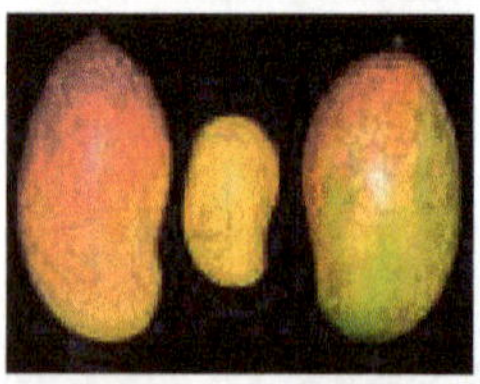	South Africa, United States	
Van Dyke		Italy, United States	
Vellai Kolomban		Sri Lanka	
Young		United States	
Zill		South Africa, United States	

List of Cultivars by Nation

- Australia: B74 (known by the brand name *Calypso*), Brooks, Green eating, Haden, Irwin, Keitt, Kensington Pride, Kent, Nam Doc Mai, Palmer, R2E2, Honey Gold.
- Bangladesh: Chok anan, Ashini, Fazli, Himsagar, Khirshapat, Langra, Lokhon-bhog, Raj-bhog
- Brazil: Coquinho, Haden, Manga Espada, Manga Rosa, Palmer, Tommy Atkins
- Cambodia: Cambodiana
- Cameroon: Améliorée du Cameroun
- China: Baiyu, Guixiang, Huangpi, Huangyu, Macheco, Sannian, Yuexi
- Costa Rica: Haden, Irwin, Keitt, Mora, Tommy Atkins
- Cuba: San Felipe, Prieto, Toledo
- Ecuador: Ambassador, Alphonso, Ataulfo, Criollos, Haden, Julie, Keitt, Kent, Reina, Tommy Atkins
- Egypt: Alphonso, Hindi, Hindi Besennara, Beid El Agl, Oweisi, Fuss Oweis, Taymoor, Zebdiah, Mesk
- Guatemala: Haden, Kent, Tommy Atkins
- Haiti: Francine (Madame Francis), Muscas, Labiche, Baptiste, Rosalie, Poirier, Corne, Fil
- Honduras: Haden, Kent, Lancetilla, Tommy Atkins
- India: Alphonso, Amrapali, Alampur Baneshan, Badshahpasand, Bangalora, Banganapalli, Black and Rose, Bombay, Bombay Green, Badami, Cheruku Rasalu, Chinna Rasalu, Chok anan, Thalimango, Chitoor, Rumani, Maharaja Pasand, Chinnarasam, Unda manfa, Dusehri, Ela Manga, Gaddamar, Gadam Mary, Fajri Kalan, Fernandian, Fajli, HusanNara, Gulabkhas, Himayath, Himsagar, Imam Pasand, Imam Hussain Mango, Jehangir, Ottu Mangai, Langra Benarsi, Pedda Rasalu, Surkha, Totapuri, Kalepad, Kishen Bhog, Komanga, Kothapalli Kobbari, Kuttiyattor, Kalami, Kesar, Lalbaug, Langra, Maldah, Malgis, Mallika, Malgoa, Mankur (Goa), Mankurad, Moovandan, Nattuma, Nannari, Neelum, Neeleshan, Panchadara Kalasa, Puliyan, Panduri Mamidi, Payri, Priyor, Rani, Rajapuri, Raspuri, Ratna, Rayal Special, Sindhooram, Safeda, Sammar Bahisht, Suvarnarekha, Totapuri Vanraj, Yahya Mariam Mango, Zardalu.
- Indonesia: Arumanis/Harumanis, Gadung/Gedong, Manalagi, Cengkir/Indramayu, Gajah, Bapang, Lalijiwo, Kueni, Golek, Kemiri, Boled, Bengkulu, Situbondo,Kelapa, Alor, Selaputih,
- Israel: Haden, Omer, Shelley, Keitt, Kent, Maya, Nimrod, Palmer, Tommy Atkins, BD-4-98
- Italy: Kensington Pride, Glenn, Tommy Atkins, Keitt, Maya, Van Dyke, Osteen, Kent
- Japan (Okinawa): Irwin

- Kenya: Apple Mango, Batwi, Boubo, Ngowe

Apple mango.

- Malaysia: Apple Mango, Apple Rumani, Arumanis, Golek, Kuala Selangor, Malgoa, Maha-65, Tok Boon
- Mali: Amelie, Kent
- Mexico: Ataulfo, Haden, Irwin, Kent, Manila, Palmer, Sensation, Tommy Atkins, Van Dyke, Petakon, Oro, Criollo, Niño,Miyako.
- Myanmar: Aug Din, Ma Chit Su, Sein Ta Lone, Shwe Hin Tha
- Nepal: Amrapali, Dusehri, Bombay, Mallika, Pharsi aamp,Supadi aamp, Sindure aamp and other local cultivars.
- Pakistan: Almaas, Alphonso, Anmol, Anwar Rataul, BaganPali, Chaunsa, Chok anan, Collector, Dusehri, Desi Ada Pamato, Desi Badam, Desi Gola, Desi Badshah, Dilkash, Fajri, Gulab Janhu, Gulab Khas, Lahoti, Lal Badshah, Langra, Maldah, Muhammad Wole, Nawab Puri, Neelum, Rani Phool, Sindhri, Saroli, Sawarnarika, Saleh Bhai, Saib, Shan-e-Khuda, Taimuria, Toofan, Wanghi, Zafran
- Peru: Criollos, Haden, Keitt, Kent, Tommy Atkins
- Philippines: Apple Mango, Carabao or Kinalabaw, Indian, Piko, Paho, Pahohutan
- Réunion island: Carotte, Jose, Lucie, Auguste
- Singapore: Apple Mango, Arumanis, Golek, Kaem Yao, Mangga Dadol
- South Africa: Fascell, Haden, Keitt, Kent, Sensation, Tommy Atkins, Zill
- Sri Lanka: Dampara, Hingurakgoda, Karutha Kolomban, Malwana[under-ground] amba, Parrot Mango and Peterpasand, Petti amba, Rata amba, Vellai Kolomban, Wild Mango, Willard, Mee Amba, Kohu Amba, Pol Amba, Giraa amba, Red Willard, Green Willard, Mallika, TJC Mango.
- Sudan: Alfonso, Bez el-Anza, Oweisi, Taymoor
- Suriname: Rood borsje,Tetéé
- Taiwan: JinHwang, Red JinHwang, Tainong No. 1, Irwin
- Tanzania: Boribo Muyini, Dodo, Mawazo, Sindano

- Thailand: Chok anan, Khaew Sawei, Nam Dok Mai, Rad, Brahm Kai Meu, Okrong
- US:
- California: Keitt, Haden, Timotayo, Manila
- Florida: Alampur Baneshan, Alice, Alphonso, Anderson, Angie, Bailey's Marvel, Bennet Alphonso, Beverly, Bombay, Brahm Kai Meu, Brooks, Carabao, Carrie, Chok Anon, Cogshall, Cushman, Dot, Duncan, Earlygold, East Indian, Edward, Eldon, Emerald, Fairchild, Fascell, Florigon, Ford, Gary, Gaylour, Glenn, Gold Nugget, Golden Lippens, Graham, Haden, Hatcher, Ice Cream, Irwin, Ivory, Jakarta, Jean Ellen, Julie, Keitt, Kensington Pride, Kent, Lancetilla, Langra Benarsi, Lippens, Mallika, Manilita, Mendoza, Mulgoba, Nam Doc Mai, Nam Tam Teen, Neelum, Nu Wun Chan, Okrung, Osteen, Palmer, Parvin, Pascual, Philippine, Pickering, Po Pyu Kalay, Rosigold, Ruby, Rutledge, Saigon, Sensation, Sophie Fry, Southern Blush, Spirit of '76, Springfels, Sunset, Suwon Tip, Tebow, Toledo, Tom Dang, Tommy Atkins, Torbert, Turpentine, Valencia Pride, Van Dyke, Zill
- Hawaii: Hawaiian Common, Gouveia, Hawaiian Dwarf, Kurahige, Mapulehu, Momi K, Pope, Rapoza, Sugai, Turpentine
- Venezuela: Haden, Keitt, Kent, Tommy Atkins
- Vietnam: Cao Lãnh Cát Chu mango, Bình Định Elephant mango, Hoà Lộc Sand mango
- West Indies: Amélie, Black (blackie), Bombay, Dou-douce, East Indian, Graham, Haden, Julie (St. Julian), Long, Madame Francis, Rose, Spice-Box, Starch

Banana Cultivars

The following is a list of banana cultivars and the groups into which they are classified. Almost all modern cultivated varieties (cultivars) of edible bananas and plantains are hybrids and polyploids of two wild, seeded banana species, *Musa acuminata* and *Musa balbisiana*. Cultivated bananas are almost always seedless (parthenocarpic) and hence sterile, so they are propagated vegetatively. They are classified into groups according to a genome-based system introduced by Ernest Cheesman, Norman Simmonds, and Ken Shepherd, which indicates the degree of genetic inheritance from the two wild parents and the number of chromosomes (ploidy). Cultivars derived from *Musa acuminata* are more likely to be used as dessert bananas, while those derived from *Musa balbisiana* and hybrids of the two are usually plantains or cooking bananas.

Left to right: plantains, Red, Latundan, and Cavendish bananas.

Classification of Cultivars

Banana plants were originally classified by Linnaeus into two species, which he called *Musa paradisiaca* – those used as cooking bananas (plantains), and *M. sapientum* – those used as dessert bananas. The primary center of diversity of cultivated bananas is Southeast Asia. Botanical exploration of this area led to many more species being named, along with subspecies and varieties. However, this approach proved inadequate to deal with the large number of cultivated varieties (cultivars) which were discovered, and many of the names later proved to be synonyms. Furthermore, it was discovered that most cultivated bananas are actually hybrids between two wild species, *M. acuminata* and *M. balbisiana*, both first described in 1820 by the Italian botanist Luigi Aloysius Colla, and that Linnaeus' two "species" were both this hybrid, which is now called *M.* × *paradisiaca*. Unlike the wild species, which have seeds, cultivated bananas are almost always seedless (parthenocarpic) and hence sterile, so they have to be propagated vegetatively.

In 1955, researchers Norman Simmonds and Ken Shepherd proposed abandoning traditional Latin-based botanical names for cultivated bananas. This approach foreshadowed the International Code of Nomenclature for Cultivated Plants which, in addition to using Latin names based on the International Code of Nomenclature for algae, fungi, and plants, gives cultivars names in a currently spoken language, enclosed in single quotes, and organizes them into "cultivar groups", also not given Latin names.

Banana cultivars derived from *M. acuminata* and *M. balbisiana* can be classified into cultivar groups using two criteria. The first is the number of chromosomes: whether the plant is diploid, triploid or tetraploid. The second is relationship to the two ancestral species, which may be determined by genetic analysis or by a scoring system devised by Simmonds and Shepherd. A cultivar is scored on 15 characters, chosen because they differ between the two species. Each character is given a score between one and five according to whether it is typical of *M. acuminata* or of *M. babisiana* or is in between. Thus the total score for a cultivar will range from 15 if all characters agree with *M. acuminata* to 75 if all characters agree with *M. balbisiana*. Intermediate scores suggest mixed ancestry: for example, 45 would be expected for diploids with equal genetic contributions from both species.

Groups are then named using a combination of the letters "A" and "B". The number of letters shows the ploidy; the proportion of As and Bs the contributions of the ancestral species. The AAB Group, for example, comprises triploid cultivars with more genetic inheritance from *M. acuminata* than *M. balbisiana*. A character score of around 35 is expected for members of this group. Within groups, cultivars may be divided into subgroups and then given a cultivar name, e.g. *Musa* AAA Group (Cavendish Subgroup) 'Robusta'.

Characters used to classify banana cultivars derived from *M. acuminata* and *M. balbisiana*		
Character	***M. acuminata***	***M. balbisiana***
Color of pseudostem	Black or grey-brown spots	Unmarked or slightly marked
Petiole canal	Erect edge, with scarred inferior leaves, not against the pseudostem	Closed edge, without leaves, against the pseudostem
Stalk	Covered with fine hair	Smooth
Pedicels	Short	Long
Ovum	Two regular rows in the locule	Four irregular rows in the locule
Elbow of the bract	Tall (< 0.28)	Short (> 0.30)
Bend of the bract	The bract wraps behind the opening	The bract raises without bending behind the opening
Form of the bract	Lance- or egg-shaped, tapering markedly after the bend	Broadly egg-shaped
Peak of the bract	Acute	Obtuse
Color of the bract	Dark red or yellow on the outside, opaque purple or yellow on the inside	Brown-purple on the outside, crimson on the inside
Discoloration	The inside of the bract is more bright toward the base	The inside of the bract is uniform
Scarification of the bract	Prominent	Not prominent
Free tepal of the male flower	Corrugated under the point	Rarely corrugated
Color of the male flower	White or cream	Pink
Color of the markings	Orange or bright yellow	Cream, yellow, or pale pink

In practice, the scoring system and the associated grouping is not as straightforward as the Simmonds and Shepherd naming system implies. For example, a member of the AAB Group should have a score about one third of the way between *M. acuminata* and *M. balbisiana* (i.e. about 35) if one third of its chromosomes come from *M. balbisiana*. However, the cultivars 'Silk' and 'Pome', both classified in the AAB Group, scored 26 and 46 respectively. The cultivar 'Pelipita' is placed in the ABB group, so should have 11 of its 33 chromosomes derived from *M. acuminata*. However, a technique called "genomic *in situ* hybridization" (GISH) showed that actually only 8 chromosomes were of this origin. Other lines of evidence suggest a more complex genome structure is present in other banana cultivars, so the group names should not be taken at face value.

Cultivars

The total number of cultivars of bananas and plantains has been estimated to be anything from around 300 to more than 1000. Names are highly confused, even within a single country. Many common names do not refer to a single cultivar or clone; for example 'Lady's Finger' or 'Lady Finger' has been used as the name for members of different genome groups, including AA and AAB. Many other names are synonyms of cultivars grown in the same or different countries. Attempts have been made to create lists of synonyms. In 2000, Valmayor et al. listed equivalent local names for 68 cultivars across five Southeast Asian countries (the Philippines, Malaysia, Indonesia, Thailand and Vietnam), together with their internationally used names. They considered a further 81 cultivars to be unique to one country. In 2007, Ploetz et al. listed more cultivar names and synonyms, with an emphasis on those grown in the islands of the Pacific, but including some grown in areas such as India, Africa and South America. As an example, for the widely grown cultivar 'Dwarf Cavendish', they gave 58 synonyms from 29 different countries or geographical areas. ProMusa has created a checklist of banana cultivar names based on available literature.

A recent development is the use of "somaclones" in banana cultivation. Micropropagation involves growing plants from very small amounts of source tissue, sometimes even a single cell, under sterile conditions using artificial techniques to induce growth. The purpose of micropropagation is often to produce a large number of genetically identical offspring. However, by inducing mutations through various means, it is possible to produce plants which differ slightly from the "parent" plant and from each other ("somaclonal variations"). By growing on these somaclones and selecting those with desirable features, new cultivars can be produced which are very similar to an existing cultivar, but differ in one or two features, such as disease resistance. Somaclones may only be distinguishable by genetic analysis.

List of Basil Cultivars

Basil cultivars are cultivated varieties of basil. They are used in a variety of ways: as culinary herbs, landscape plants, healing herbs, teas, and worship implements. All true basils are species of the genus *Ocimum*. The genus is particularly diverse, and includes annuals, non-woody perennials and shrubs native to Africa and other tropical and subtropical regions of the Old and New World. Although it is estimated that there are 50 to 150 species of basil, most, but not all, culinary basils are cultivars of *O. basilicum*, or sweet basil. Some are cultivars of other basil species, and others are hybrids. It is particularly challenging to determine which species a basil belongs to. This is because basil cross-breeds easily, and drawing boundaries between species is particularly difficult. In fact, recent studies have led to reclassification of some portions of the genus.

Basil cultivars vary in several ways. Visually, the size and shape of the leaves varies greatly, from the large lettuce-like leaves of the *Mammoth basil* and *Lettuce leaf basil* to the tiny leaves of the *Dwarf bush basil*. More practically, the fragrance of the basil varies due to the varying types and quantities of essential oils contained in the plants. The most important are 1,8 cineol, linalool, citral, methyl chavicol (estragole), eugenol and methyl cinnamate, although hardly any basil contains all of these in any significant amount.

Basil Cultivars

Holy basil.

Sweet basil growing in the sun.

Late summer Thai basil flowers.

Harvested sweet basil leaves.

Greek Bush Basil.

***Ocimum basilicum* cultivars**		
Common name	**Species and cultivars**	**Description**
Sweet basil	*O. basilicum*	With a strong clove scent when fresh.
Lettuce leaf basil	*O. basilicum* 'Lettuce Leaf'	Has leaves so large they are sometimes used in salads.
Mammoth basil	*O. basilicum* 'Mammoth'	Another large-leaf variety, stronger flavor than sweet Genovese.
Genovese basil	*O. basilicum* 'Genovese Gigante'	Almost as popular as sweet basil, with similar flavor.
Nufar basil	*O. basilicum* 'Nufar F1'	Variety of Genovese resistant to fusarium wilt.
Spicy globe basil	*O. basilicum* 'Spicy Globe'	Grows in a bush form, very small leaves, strong flavor.

Greek Yevani basil	*O. basilicum* 'Greek Yevani'	Organically grown version of Spicy globe basil.
Fino verde basil	*O. basilicum piccolo*	Small, narrow leaves, sweeter, less pungent smell than larger leaved varieties.
Boxwood basil	*O. basilicum* 'Boxwood'	Grows tightly like boxwood, very small leaves, strong flavor, great for pestos.
Purple ruffles basil	*O. basilicum* 'Purple Ruffles'	Solid purple, rich and spicy and a little more anise-like than the flavor of Genovese Basil.
Magical Michael	*O. basilicum* 'Magical Michael'	Award-winning hybrid with an uncommon degree of uniformity, and nice flavor for culinary use.
Dark opal basil	*O. basilicum* 'Purpurascens'	Award-winning variety, developed at the University of Connecticut in the 1950s.
Red rubin basil	*O. basilicum* 'Red Rubin'	Strong magenta color, similar flavor to sweet basil, also called *Opal basil.*
Osmin purple basil	*O. basilicum* 'Osmin Purple'	Dark shiny purple with a jagged edge on the leaves, smaller leaves than red rubin.
Cuban basil	*O. basilicum*	Similar to sweet basil, with smaller leaves and stronger flavor, grown from cuttings.
Thai basil	*O. basilicum var. thyrsiflorum*	Called Ho-ra-pa (ต้นโหระพา) in Thai, gets its scent of licorice from estragole.
'Siam Queen'	*O. basilicum var. thyrsiflorum* 'Siam Queen'	A named cultivar of Thai Basil
Cinnamon basil	*O. basilicum* 'Cinnamon'	Also called *Mexican spice basil*, with a strong scent of cinnamate, the same chemical as in cinnamon. Has purple flowers.
Licorice basil	*O. basilicum* 'Licorice'	Also known as *Anise basil* or *Persian basil*, silvery leaves, spicy licorice smell comes from the same chemical as in anise, anethole. Thai basil is also sometimes called *Licorice basil.*
Mrs. Burns lemon basil	*O. basilicum* var. *citriodora* 'Mrs. Burns'	Clean, aromatic lemon scent, similar to lemon basil.
***Ocimum americanum* (formerly known as *O. canum*) cultivars**		
Common name	**Species and cultivars**	**Description**

Lemon basil	*O. americanum*	Contains citral and limonene, therefore actually does smell very lemony, tastes sweeter. Originally, and sometimes still, called "hoary basil". Popular in Indonesia, where it is known as 'kemangi'. Also sometimes 'Indonesian basil'.
Lime basil	*O. americanum*	Similar to lemon basil.
***Ocimum ×citriodorum* cultivars**		
Common name	**Species and cultivars**	**Description**
Greek column basil	*O. ×citriodorum* 'Lesbos'	Columnar basil, can only be propagated from cuttings.
Thai lemon basil	*O. ×citriodorum*	Called mangluk (แมงลัก) in Thai. It has a citrus odor, with a distinct Lemon-balm-like flavor
Other Species and Hybrids		
Common name	**Species and cultivars**	**Description**
Holy basil	*O. sanctum* (alt. *O. tenuiflorum*)	Also *sacred basil*, *Tulsi* (तुलसी) in Hindi, a perennial breed from India, used in Ayurveda, for worship, and in Thai cooking.
Clove Basil	*O. gratissimum*	
Greek bush basil (Greek Spicy Globe Basil)	*O. minimum* (alt. *O. basilicum* var. *minimum*)	Forms a nearly perfectly round globe, with thin, tiny leaves and a delicious scent. Despite its name, the variety probably originated in Chile.
Dwarf bush basil	*O. minimum*	Unusually small bush variety, similar to *Greek bush basil*.
African Blue basil	*O. kilimandscharicum × basilicum*	A sterile perennial hybrid, with purple coloration on its leaves and containing a strong portion of actual camphor in its scent.
Spice basil	*O. basilicum × americanum*	A fruity/musky-scented cultivar sometimes sold as Holy Basil
Sweet Dani basil	*O. basilicum × americanum*	A vigorous, large-leaved green basil with a strong, fresh lemon scent, a 1998 All-American Selection.

List of *Canna* Cultivars

This list of *Canna* cultivars is a gallery of named cultivars of plants in the genus *Canna* that are representative of the various *Canna* cultivar groups (i.e., groups of very similar cultivars).

Names of cultivars conform to the rules of the International Society for Horticultural Science (ISHS) Commission for Nomenclature and Cultivar Registration, as laid down in the International Code of Nomenclature for Cultivated Plants. They are registered with an International Cultivar Registration Authority (ICRA), which for the *Canna* genus is the Royal General Bulbgrowers' Association of the Netherlands (KAVB).

Foliage Group

Cultivars, F1 and F2 hybrids, normally with small species-like flowers, but grown principally for their foliage. This group has occasionally been referred to as the Année Group, after the originator, Théodore Année, the world's first *Canna* hybridizer. However, the use of an accented character in the name creates problems, both in pronunciation and keyboard entry, that it was felt that as they were grown primarily for foliage, then "Foliage Group" was the better name choice.

Canna 'Argentina', I. Cooke 1999

Canna 'Auguste Ferrier'

Canna (Agriculture Group) 'Edulis Dark'

Canna 'Ferrandii'

Canna 'Musaefolia Grande', H. Kelly Jnr.

Crozy Group

A cultivar group where the flower spikes are arranged close together on the stalk and have narrow to medium petals. There is always space between the staminodes when arranged formally, and the labellum (lip) is smaller than the staminodes, and is often twisted or curled.

The pioneer of this group was Monsieur Crozy of Lyons, France, who started breeding cannas as early as 1862, from stock originally developed by Théodore Année.

They are sometimes referred to as gladiolus flowering cannas, but describing flowers as similar to another genus is not encouraged. In the past, they were called the *Canna* × *generalis* L.H. Bailey garden species, but the use of nothospecies for complex garden hybrids have been replaced by Cultivar Groups in the International Code of Nomenclature for Cultivated Plants.

Canna 'R. Wallace', Pfitzer 1902

Canna 'Anette Dalebö', Dalebö 2006

Canna 'Antonin Crozy', Crozy

Canna 'Apricot Dream', K. Kelly

Canna 'Florence Vaughan', Crozy 1892

Canna 'Jessie Dalebö', Dalebö 2003

Italian Group

A cultivar group with large, fragile staminodes. Flowers are arranged somewhat loosely, with wide petals, so wide that there is no space between them, when arranged formally. The labellum (lip) is larger, or at least as large, as the staminodes, unlike the other groups where it is smaller and sometimes curled. The stamen is also much wider than that in the other cultivar groups.

Also, used to be called the orchid flowering cannas, or *C.* × *orchiodes* L.H. Bailey garden species, although such "pretend" species are now deprecated in favor of Cultivar Groups. In any event, it is difficult to see the similarity between this group and orchids.

Most of this group obtained its larger sized flowers from the introduction of *Canna flaccida* in the early 1890s by Dr Sprenger in Naples, Italy followed shortly afterwards by Luther Burbank in California, USA, with the same cross.

Canna 'Allemania', Sprenger 1896

Canna 'American Flag', Wintzer 1902

Canna 'Austria', Sprenger 1893

Canna 'Bavaria', Sprenger 1895

Canna 'Burbank', Burbank 1895

Canna 'Burgundia', Sprenger 1896

Canna 'Italia', Sprenger 1893

Australian Group

The result of a crossing of a Foliage Group seed parent with an Italian Group pollen parent (*C.* 'Red Stripe' x *C.* 'Bengal Tiger'), which resulted in the very large leaves of the Foliage Group allied with large flowers.

Canna 'Polanaise', Bernard Yorke

Premier Group

This grouping contains cultivars that have a large, circular shape, without gaps between the staminodes when ordered. These are derived from triploids and crosses with the Italian Group cultivars.

Canna 'Aida', Howard & Smith 1920's

Canna 'Alfred Young', Dalebö 2005

Canna 'Eureka'

Canna 'Karla Dalebö', Dalebö 2007

Variegated Group

Cultivars with variegated foliage, regardless of what other Group they may belong to.

Canna (Italian Group) 'Bengal Tiger'

Canna (Italian Group) 'Phasion'

Canna (Italian Group) (Miniature Group) 'Pringle Bay', Bloch

Canna (Italian Group) 'Trinacria Variegata'

Conservatory Group

The growing conditions in a conservatory are quite specialized and do not suit many cultivars, this group have been selected for thriving in this environment, required features being plant vigor, early flowering, foliar appearance, self-cleaning ability and good propagation qualities. The originator of this group was Robert Armstrong (geneticist) while he was working at Longwood Gardens in the USA in 1967.

Canna 'Chesapeake', Armstrong

Canna 'Constitution', Armstrong

Canna 'Lenape', Armstrong

Aquatic Group

Cultivars that thrive as marginal water plants. Characteristically, they will have lance-shaped foliage and long, thin rhizomes.

Canna 'Endeavour', Armstrong

Canna 'Erebus', Armstrong

Canna 'Erebus', Armstrong

Miniature Group

Cultivars growing under 0.5m (19") high, the flowers should be in scale to the rest of the plant.

Canna 'Puck'

Agriculture Group

Cultivars grown selectively for agricultural usage, normally for their very large rhizomes and a high starch yield.

Canna 'Achira Dark'

Canna 'Queensland Arrowroot'

Musaefolia Group

The Musaefolia Group, having large leaves and being either tall or giant in size consists of cultivars whose leaves resemble those of banana plants (genus *Musa*). Until this group was designated, the cultivars were considered to be members of the Foliage Group.

Canna 'Musaefolia Hybrida'

Canna 'Musaefolia Hybrida'

Canna 'Musaefolia Peruviana'

Citrus

Citrus is a genus of flowering trees and shrubs in the rue family, Rutaceae. Plants in the genus produce **citrus fruits**, including important crops like oranges, lemons, grapefruit, pomelo and limes.

The most recent research indicates an origin in Australia, New Caledonia and New Guinea. Some researchers believe that the origin is in the part of Southeast Asia bordered by Northeast India, Burma (Myanmar) and the Yunnan province of China, and it is in this region that some commercial species such as oranges, mandarins, and lemons originated. Citrus fruit has been cultivated in an ever-widening area since ancient times.

Cultivation

Citrus trees hybridise very readily – depending on the pollen source, plants grown from a Persian lime's seeds can produce fruit similar to grapefruit. Thus all commercial citrus cultivation uses trees produced by grafting the desired fruiting cultivars onto rootstocks selected for disease resistance and hardiness.

Lemons are a citrus fruit native to Asia, but now common worldwide.

Persian limes in a grocery store.

The color of citrus fruits only develops in climates with a (diurnal) cool winter. In tropical regions with no winter at all, citrus fruits remain green until maturity, hence the tropical "green oranges". The Persian lime in particular is extremely sensitive to cool conditions, thus it is not usually exposed to cool enough conditions to develop a mature color. If they are left in a cool place over winter, the fruits will change color to yellow.

The terms "ripe" and "mature" are usually used synonymously, but they mean different things. A mature fruit is one that has completed its growth phase. Ripening is the changes that occur within the fruit after it is mature to the beginning of decay. These changes usually involve starches converting to sugars, a decrease in acids and a softening and change in the fruit's color.

Citrus fruits are non-climacteric and respiration slowly declines and the production and release of ethylene is gradual. The fruits do not go through a ripening process in the sense that they become "tree ripe." Some fruits, for example cherries, physically mature and then continue to ripen on the tree. Other fruits, like pears, are picked when mature but before they ripen, then continue to ripen off the tree. Citrus fruits pass from immaturity to maturity to over-maturity while still on the tree. Once they are separated from the tree, they will not increase in sweetness or continue to ripen. The only way change may happen after being picked is that they will eventually start to decay.

With oranges, color cannot be used as an indicator of ripeness because sometimes the rinds turn orange long before the oranges are ready to eat. Tasting them is the only way to know whether or not they are ready to eat.

Mediterranean Mandarin (*Citrus ×deliciosa* plantation, Son Carrió (Mallorca).

Citrus trees are not generally frost hardy. Mandarin oranges (*C. reticulata*) tend to be the hardiest of the common *Citrus* species and can withstand short periods down to as cold as −10 °C (14 °F), but realistically temperatures not falling below −2 °C (28 °F) are required for successful cultivation. Tangerines, tangors and yuzu can be grown outside even in regions with more marked sub-freezing temperatures in winter, although this may affect fruit quality. A few hardy hybrids can withstand temperatures well below freezing, but do not produce quality fruit. Lemons can be commercially grown in cooler-summer/moderate-winter, coastal Southern California, because sweetness is neither attained nor expected in retail lemon fruit. The related trifoliate orange (*Citrus trifoliata*) can survive below −20 °C (−4 °F); its fruit are astringent and inedible unless cooked but a few better-tasting cultivars and hybrids have been developed.

Leaf of Citrus tree.

The trees thrive in a consistently sunny, humid environment with fertile soil and adequate rainfall or irrigation. Abandoned trees in valleys may suffer, yet survive, the dry summer of Central California's Inner Coast Ranges. At any age citrus grows well enough with infrequent irrigation in partial shade, but the fruit crop is smaller. Being of tropical and sub-tropical origin, oranges, like all, citrus are broadleaved and evergreen. They do not drop leaves except when stressed. The stems of many varieties have large sharp thorns. The trees flower in the spring, and fruit is set shortly afterward. Fruit begins to ripen in fall or early winter months, depending on cultivar, and develops increasing sweetness afterward. Some cultivars of tangerines ripen by winter. Some, such as the grapefruit, may take up to eighteen months to ripen.

Production

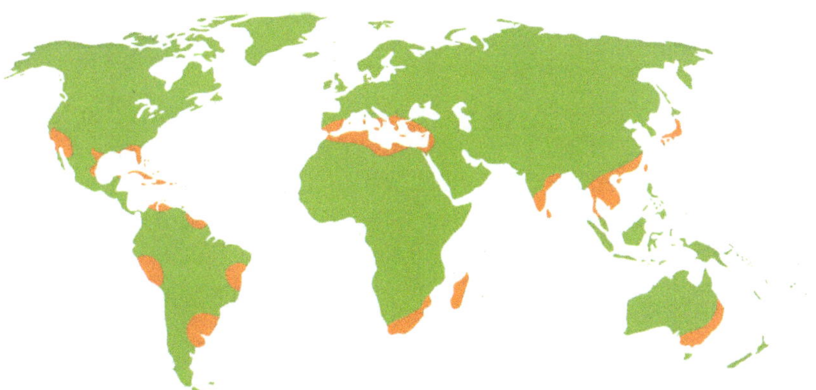

Major producer regions.

According to UN 2007 data, Brazil, China, the United States, Mexico, India, and Spain are the world's largest citrus-producing countries.

Major commercial citrus growing areas include southern China, the Mediterranean Basin (including southern Spain), South Africa, Australia, the southernmost United States, Mexico and parts of South America. In the United States, Florida, California, Arizona, and Texas are major producers, while smaller plantings are present in other Sun Belt states and in Hawaii.

As Ornamental Plants

Citrus trees grown in tubs and wintered under cover were a feature of Renaissance gardens, once glass-making technology enabled sufficient expanses of clear glass to be produced. An orangery was a feature of royal and aristocratic residences through the 17th and 18th centuries. The *Orangerie* at the Palace of the Louvre, 1617, inspired imitations that were not eclipsed until the development of the modern greenhouse in the 1840s. In the United States the earliest surviving orangery is at the Tayloe House, Mount Airy, Virginia. George Washington had an orangery at Mount Vernon.

Orangery of the Botanical Garden in Leuven (Belgium).

Some modern hobbyists still grow dwarf citrus in containers or greenhouses in areas where it is too cold to grow it outdoors. Consistent climate, sufficient sunlight, and proper watering are crucial if the trees are to thrive and produce fruit. Compared to many of the usual "green shrubs", citrus trees better tolerate poor container care. For cooler winter areas, limes and lemons should not be grown, since they are more sensitive to winter cold than other citrus fruits. Hybrids with kumquats (× *Citrofortunella*) have good cold resistance.

Pests and Diseases

Citrus plants are very liable to infestation by aphids, whitefly and scale insects (e.g. California red scale). Also rather important are the viral infections to which some of these ectoparasites serve as vectors such as the aphid-transmitted *Citrus tristeza virus* which when unchecked by proper methods of control is devastating to citrine plantations. The newest threat to citrus groves in the United States is the Asian citrus psyllid.

The Asian citrus psyllid is an aphid-like insect that feeds on the leaves and stems of citrus trees and other citrus-like plants – but the real danger lies in that it can carry a deadly, bacterial tree disease called Huanglongbing (HLB), also known as citrus greening disease.

Citrus canker is caused by the gammaproteobacterium *Xanthomonas axonopodis.*

In August 2005 citrus greening disease was discovered in the south Florida region around Homestead and Florida City. It has since spread to every commercial citrus grove in Florida. In 2004-2005 USDA Ag statistics reported the total Florida citrus production to be 169.1 million boxes of fruit. The estimate for all Florida citrus production in the 2015-2016 season is 86.4 million boxes—a 51% drop.

In June 2008, the psyllid was spotted dangerously close to California – right across the international border in Tijuana, Mexico. Only a few months later, it was detected in San Diego and Imperial counties, and has since spread to Riverside, San Bernardino, Orange, Los Angeles and Ventura counties sparking quarantines in those areas. The Asian citrus psyllid has also been intercepted coming into California in packages of fruit and plants, including citrus, ornamentals, herbs and bouquets of cut flowers, shipped from other states and countries.

The foliage is also used as a food plant by the larvae of Lepidoptera (butterfly and moth) species such as the Geometridae common emerald (*Hemithea aestivaria*) and double-striped pug (*Gymnoscelis rufifasciata*), the Arctiidae giant leopard moth (*Hypercompe scribonia*), *H. eridanus*, *H. icasia* and *H. indecisa*, many species in the family Papilionidae (swallowtail butterflies), and the black-lyre leafroller moth (*"Cnephasia" jactatana*), a tortrix moth.

Since 2000, the citrus leafminer (*Phyllocnistis citrella*) has been a pest in California, boring meandering patterns through leaves.

In eastern Australia, the bronze-orange bug (*Musgraveia sulciventris*) can be a major pest of citrus trees, particularly grapefruit. In heavy infestations it can cause flower and fruit drop and general tree stress.

European brown snails (*Cornu aspersum*) can be a problem in California, though laying female Khaki Campbell and other mallard-related ducks can be used for control.

Deficiency Diseases

Citrus plants can also develop a deficiency condition called chlorosis, characterized by yellowing leaves highlighted by contrasting leaf veins. The shriveling leaves eventually fall, and if the plant loses too many, it will slowly die. This condition is often caused by an excessively high pH (alkaline soil), which prevents the plant from absorbing iron, magnesium, zinc, or other nutrients it needs to produce chlorophyll. This condition can be cured by adding an appropriate acidic fertilizer formulated for citrus, which can sometimes revive a plant to produce new leaves and even flower buds within a few weeks under optimum conditions. A soil which is too acidic can also cause problems; citrus prefers neutral soil (pH between 6 and 8). Citrus plants are also sensitive to excessive salt in the soil. Soil testing may be necessary to properly diagnose nutrient deficiency diseases.

List of Coffee Varieties

Coffee varieties are the diverse subspecies derived through selective breeding or natural selection of coffee plants. While there is tremendous variability encountered in both wild and cultivated coffee plants, there are a few varieties and cultivars that are commercially important due to various unique and inherent traits such as disease resistance and fruit yield. These unique traits are what producers use to select breeds when developing crops. Therefore, at a micro level, breed selection is critical to the success of a producer and is one of the key components of cup quality.

At a macro level, the viability of the coffee industry as a whole is dependent upon breed selection. Already, the majority of coffee produced originates from producers using selected breeds. For this reason, breed selection is an important aspect of sustainability within coffee production.

Terminology

There is considerable confusion as to which term to use when speaking about coffee subspecies. For the sake of clarity, within this article the terms will be used in accordance with loose guidelines put forth by the Specialty Coffee Association of America:

Variety: This rank of taxa delineates differences between plants that are smaller than in subspecies but larger than forms. A variety retains most of the characteristics of the species, but differs in some way. Cultivar: Any variety produced by horticultural or agricultural techniques and not normally found in natural populations; a cultivated variety. Most of the varieties we know in specialty coffee are really cultivars. Bourbon and Typica are some of the most widely known cultivars.

Put simply: In this article, varieties are naturally occurring subspecies and cultivars are cultivated subspecies. In addition, a third term, "breed" will be used as an umbrella term to simplify discussions in which the nuances between the terms 'variety' and 'cultivar' have no bearing.

Selection Criteria

Farmers have designed standards for crops they would cultivate. Recent advances in breeding techniques have provided farmers with higher-yielding breeds with better disease resistance and better cup quality — all traits critical to the success of a producer. Below are some additional traits a producer may use to select breeds for crop development.

Yield is the measure of the amount of fruit produced by a given breed. It is usually expressed as kilograms or tonnes per hectare per year, assuming conventional plant densities of 1,100 - 1,400 trees per hectare. High yield is one of the prime objectives of producers and breeding specifically to achieve higher yield is a relatively new trend.

Resistance to diseases has become a dominant factor in not only breed selection but also of breeding new cultivars. Breeding for disease resistance has been mainly restricted to CLR and coffee berry disease (CBD) but also includes other diseases of local importance.

Resistance to Pests is not normally a trait developed by breeding but is rather a trait selected from among breeds. Certain breeds of coffee have been found to be resistant to nematodes and leaf miner. As with diseases, robusta has been found to be the more resistant species compared to arabica.

Arabica Varieties and Cultivars

According to The International Trade Centre, Arabica coffee accounted for roughly 61 per cent of the world's coffee production between 2004 - 2010. It would be higher if Arabica were not as susceptible to disease as it is. Coffee from the species *C. arabica* has many different varieties, each with unique characteristics. Some well-known arabica coffees include:

Name	Species	Region(s)	Comments	Ref
Arusha	Arabica	Mount Meru in Tanzania, and Papua New Guinea	Typica variety or a French Mission.	
Bergendal, Sidikalang	Arabica	Indonesia	Typica varieties which survived the leaf rust outbreak of the 1880s; most of the other Typica in Indonesia was destroyed.	
Blue Mountain	Arabica	Blue Mountains region of Jamaica. Also grown in Kenya, Hawaii, Haiti, Papua New Guinea (where it is known as PNG Gold) and Cameroon (where it is known as Boyo).	Typica, known to have some resistance to coffee berry disease.	

Bourbon	Arabica	Réunion, Rwanda, Latin America.	Around 1708 the French planted coffee on the island of Bourbon (now called Réunion) in the middle of the Indian Ocean, all probably from the same parent stock – the plant the Dutch gave them. Unsurprisingly, it mutated slightly and was planted throughout Brazil in the late 1800s and eventually spread through Latin America. Bourbon produces 20–30% more fruit than Typica|Typica varieties. El Salvador is known as the Bourbon Country.	
Catuai	Arabica	Latin America	This is a hybrid of Mundo Novo and Caturra bred in Brazil in the late 1940s.	
Catimor	Interspecific hybrid	Latin America, Indonesia	This is cross between Timor coffee and Caturra coffee. It was created in Portugal in 1959.	
Caturra	Arabica	Latin and Central America	This coffee varietal was developed by the Alcides Carvalho Coffee Center (Centro de Cafe) of the IAC, Instituto Agronomico of Campinas, São Paulo State in Brazil. In 1937, IAC received seed samples of genetic materials originated on the border of the states of Minas Gerais and Espírito Santo. It was from Red Caturra and yellow Caturra cultivars. These two cultivars originated by natural mutation of Bourbon Red, originally a tall coffee shrub, found in the Serra do Caparaó, which is now a mountainous National Park north east of the city of Rio de Janeiro. It produces a higher yield than Bourbon, and this is generally due to the plant being shorter and with less distance between the branches. A relatively recently selected botanical variety of the Coffea arabica species that generally matures more quickly, produces more coffee, and is more disease resistant than older, traditional arabica varieties. In fact this mutation is not unique; it led to the formation of the Pacas variety in El Salvador (from Bourbon) and the Villa Sarchi in Costa Rica (from Bourbon). Genetically it is very similar to Bourbon although it usually produces a poorer cup quality but this is mainly due to the variety yielding more.	

Charrieriana	Not yet named	Cameroon	This is a newly found variety from Cameroon. It has gained some press recently due to its caffeine-free nature. Not yet grown commercially, but it probably will be.	
Colombian	Arabica	Colombia	Coffee was first introduced to the country of Colombia in the early 1800s. Today Maragogipe, Caturra, Typica|Typica and Bourbon cultivars are grown. When Colombian coffee is freshly roasted it has a bright acidity, is heavy in body and is intensely aromatic. Colombia accounts for about 12% of the coffee market (by value) in the world, third in volume after Vietnam and Brazil.	
Ethiopian Harar	Arabica	Ethiopia	From the region of Harar, Ethiopia. Known for its complex, fruity flavor that resembles a dry red wine. All three Ethiopian varieties are trademarked names with the rights owned by Ethiopia.	
Ethiopian Sidamo	Arabica	Ethiopia	From the Sidamo (now Oromia) region of Ethiopia as well. All three Ethiopian varieties are trademarked names with the rights owned by Ethiopia.	
Ethiopian Yirgacheffe	Arabica	Ethiopia	From the Yirgachefe district in the Gedeo Zone of the Southern Nations, Nationalities, and People's Region of Ethiopia. All three Ethiopian varieties are trademarked names with the rights owned by Ethiopia.	
French Mission	Arabica	Africa	French Mission is actually Bourbon that was planted in East Africa by French Missionaries around 1897.	
Gesha / Geisha T.2722	Arabica	Ethiopia, Tanzania, Costa Rica, Panama, Colombia, Peru	Gesha or Geisha variety, grown in the highlands of Boquete in Chiriquí Province, Panama, highly sought after at auction, achieving high prices. Originally from the village of Gesha, Ethiopia. It was planted in the 1950s as a rust-resistant crop and rediscovered in the early 2000s. The most expensive varietal at coffee auctions, fetching $350.25USD in 2013.	
Guadeloupe Bonifieur	Arabica	Guadeloupe		
Hawaiian Kona	Arabica	Hawaii	Grown on the slopes of Hualalai and Mauna Loa in the Kona District on the Big Island of Hawaii. Coffee was first introduced to the Islands by Chief Boki, the Governor of Oahu, in 1825.	

Java	Arabica, Robusta and interspecific hybrids	Indonesia	From the island of Java, in Indonesia. This coffee was once so widely traded that "java" became a slang term for coffee. Java encompasses a regional style, not a cultivar of coffee.	
K7	Arabica	Africa	A Kenyan selection of French Mission Bourbon selected at Legelet Estate in Muhoroni, Kenya. Selected based on cupping trials.	
Maragogipe	Arabica	Latin America	Typica. It was first discovered near Maragogipe, in Brazil's state Bahia. Maragogype is well known for producing big beans.	
Maragaturra	Arabica	Latin America	Maragaturra is a man-made hybrid plant between Caturra and Maragogype.|It was first bred in order to capture the flavor profile of Maragogype with the higher yield and efficiency of the Caturra Varietal.	
Mayagüez	Arabica	Africa	A Bourbon cultivar grown in Rwanda.	
Mocha	Arabica	Yemen	Yemeni coffee traded through the once major port of Mocha. Not to be confused with the preparation style (coffee with cocoa).	
Mundo Novo	Arabica	Latin America	Mundo Novo is a hybrid between Bourbon and Typica|Typica, crossed in the 1940s.	
Orange, Yellow Bourbon	Arabica	Latin America	Red Bourbon and Orange Bourbon are types of Bourbon that have been selected from spontaneous mutation.	
Pacamara	Arabica	Latin America	Pacamara is a hybrid between the Bourbon mutation Pacas and Maragogype. It was bred in El Salvador in 1958 probably to achieve a Typica|Typica variety that produces larger beans.	
Pacas	Arabica	Latin America	A natural mutation of the Bourbon variety found in El Salvador in 1949.	
Pache Colis	Arabica	Latin America	Pache Colis is a hybrid between Pache Comum and Caturra. This variety produces distinctly larger fruit and roughly textured foliage.	
Pache Comum	Arabica	Latin America	Typica first found in Santa Rosa, Guatemala.	
Ruiru 11	Arabica	Kenya	Ruiru 11 was released in 1985 by the Kenyan Coffee Research Station. While the variety is generally disease resistant, it produces a lower cup quality than K7, SL28 and 34.	

S795	Arabica	India, Indonesia	Probably the most commonly planted Arabica in India and Southeast Asia, known for its balanced cup and subtle flavour notes of mocca. Released during the 1940s, it is a cross between the Kents and S.288 varieties.	
Santos	Arabica	Brazil	Brazil Santos is usually used as a grading term for Brazilian coffee rather than a variety of Arabica. The name refers to the port in Brazil where coffee passed through, and was regarded as higher quality than "Brazilian coffee". Brazilian Santos is usually of the Bourbon variety.	
Sarchimor	Interspecific hybrid	Costa Rica, India	A hybrid between the Costa Rican Villa Sarchi and the Timor variety. Because of its Timor parent, Sarchimor is quite resistant to leaf rust disease and stem borer. As well as Costa Rica, it is grown in India.	
SL28	Arabica	Kenya	A selection, by Scott Labs in Kenya from the Tanganyika Drought Resistant variety from northern Tanzania in 1931. Excellent flavour, commonly blackcurrant acidity.	
SL34	Arabica	Kenya	Selected by Scott Labs from the French Mission variety grown in Kenya. Selected for its superior cup quality (although inferior to SL28), but not resistant to CBD, CLR or BBC.	
Sulawesi Toraja Kalossi	Arabica	Indonesia	Actually the S795 varietal, grown at high altitudes on the island of Sulawesi (formerly Celebes), Indonesia. Kalossi is the small town in central Sulawesi which serves as the collection point for the coffee and Toraja is the mountainous area in which the coffee is grown. Sulawesi exhibits a rich, full body, well-balanced acidity and is multi-dimensional in character. Sulawesi itself is not a cultivar of coffee.	
Sumatra Mandheling and Sumatra Lintong	Arabica	Indonesia	Mandheling is named after the Mandailing people located in North Sumatra, Indonesia. The name is the result of a misunderstanding by the first foreign purchaser of the variety, and no coffee is actually produced in the "Mandailing region". Lintong on the other hand, is named after the Lintong district, also located in North Sumatra. This is not a specific cultivar, but rather a region with a specific processing style.	

Timor, Arabusta	Interspecific hybrid	Indonesia	Timor is not actually a variety of coffea arabica, but a hybrid of two species of coffee; coffea arabica and coffea canephora (also called Robusta). It was found on the island of Timor around the 1940s and it was cultivated because of its resistance to leaf rust (which most arabica coffee is susceptible to). It is called Hybrido de Timor in the Americas and Tim Tim or Bor Bor in Indonesia. Another hybrid between the two species is called Arabusta but generally only found in Africa.	
Typica	Arabica	Worldwide	Typica originated from Yemeni stock, taken first to Malabar, India, and later to Indonesia by the Dutch. It later made its way to the West Indies to the French colony at Martinique. Typica has genetically evolved to produce new characteristics, often considered new varietals: Criollo (South America), Arabigo (Americas), Kona (Hawaii), Pluma Hidalgo (Mexico), Garundang (Sumatra), Blue Mountain (Jamaica, Papua New Guinea), San Bernado & San Ramon (Brazil), Kents & Chickumalgu (India)	
Uganda	Arabica/ Robusta		Although it mostly produces Robusta coffee, there is a quality Arabica bean grown there known as Bugishu around the Sipi Falls area.	
Brutte	Arabica		Variety of coffee (arabica) Bred in 2014 in the south of India in g.Madras, 1996 Chennai Tamil Nadu. Grown at an altitude of 1500 m above sea level, which in itself is a good indicator. Differ by more quantitative tannin to 14 -15% and trigonelline 1.5 - 1.7%.	

Robusta Varieties

Vietnam is the world's largest Robusta producer, with Robusta accounting for 97% of Vietnam's coffee output.

While not separate varieties of bean, unusual and very expensive robustas are the Indonesian *Kopi Luwak* and the Philippine "Kapéng Alamid". The beans are collected from the droppings of the common palm civet, whose digestive processes give it a distinctive flavor.

Unroasted coffee beans of the Robusta variety (*Coffea canephora*).

Other Varieties

Although not as commercially viable as Arabica or Robusta, other species of coffee exist and contain unique varieties. These include *Kapeng barako or Café Baraco*, (English: Barako coffee), a *Liberica*. variety grown in the Philippines, particularly in the provinces of Batangas and Cavite. It was introduced during the period of Spanish colonization and the Philippines quickly rose to become the 4th largest producer of coffee in the early 19th century. However, production was cut short due to "coffee rust" infestation.

List of *Narcissus* Horticultural Divisions

The *Narcissus* horticultural divisions are system of classifying the cultivated varieties of the genus *Narcissus* , which are predominantly spring perennial plants in the Amaryllidaceae (amaryllis) family. Various common names including daffodil, narcissus, and jonquil are used to describe all or some members of the genus.

The list of *Narcissus* horticultural divisions provided by the Royal Horticultural Society (RHS) is the standard method used to classify and describe cultivated varieties (cultivars) of *Narcissus*. It is widely used since the RHS is the international authority for the registration of such cultivars.For horticultural purposes, all *Narcissus* cultivars are split into 13 divisions, as described by Kington (1998), for the RHS, based partly upon flower form (shape and length of corona – the "trumpet" or "cup"), particularly the ratio of corona to length of perianth segments (tepals or "petals"), the number of flowers per stem, flowering period and partly upon the genetic background. Division 11 (Split-corona) with its two subdivisions was the most recent group to be described (1969). Division 13, which includes all the wild rather than cultivated daffodils, is the exception to this scheme. This classification is a useful tool for planning planting. Most commercially available narcissi come from Divisions 1 (Trumpet), 2 (Large-cupped) and 8 (Tazetta).

Range of *Narcissus* cultivars.

Growers register new daffodil cultivars by name and color with the RHS, whose *International Daffodil Register* is regularly updated with supplements available online and is searchable. The most recent supplement (2014) is the sixth; the fifth was published in 2012. More than 27,000 names were registered as of 2008, and the number has continued to grow. However, because of synonymy, the actual number is probably closer to 18,000; only about 500 are in commercial production (470 in 2009–2010).

Registered daffodils are given a division number and color code such as 5W-W ('Thalia'). In horticultural usage it is not uncommon to also find another unofficial division of 'Miniatures', which, although drawn from the other 13 divisions, have their miniature size in common. These are sometimes referred to by nurseries as 'Division 14'. Over 140 narcissus cultivars have gained the Royal Horticultural Society's Award of Garden Merit.

Color Code

Daffodils may be self-colored—i.e., both perianth and corona identical in color and shade—or the colors between the perianth and corona may differ widely. Some perianths and some coronas may also contain more than one color or shade. Prevalent colors are all shades and tones of yellow, white, orange, pink, red and green. Pinks vary from apricot to rose in shades from pale to deep, and some more recent cultivars have hints of lavender or lilac. Reds vary from orange-red to salmon red to near scarlet. Pink, red, orange and green tones are mainly confined to the corona. However, breeders are currently working against the genus' natural pigmentation and genetic barriers to create cultivars in which pink, rose, red, orange and green tones suffuse or "bleed" from the more highly colored coronas onto the perianth segments of white or yellow. There is an increasing number of commercially available varieties which display this enhanced coloration.

Narcissus 'Geranium' **8W-O**

The color classification lists the perianth color and then the corona color. In the case of multiple colors, the perianth colors are assigned from the outer edge of the perianth segments inward to their juncture with the base of the corona, while the corona colors are assigned from the base of the corona outward to the rim. Thus, 'Accent', a Large Cup (Division 2) Daffodil possessing a white perianth and a pink corona, is officially classified as 2W-P, 'Geranium', Tazetta (Division 8) with a white perianth and orange corona is 8W-O and 'Actaea', Poeticus (Division 9), with white corona and multicolored corona is 9W-GYR.

RHS color classification	
Code	**Color**
W	White or Whitish
G	Green
Y	Yellow
P	Pink
O	Orange
R	Red

Definitions of Divisions

RHS horticultural divisions of daffodils			
Number	**Name**	**Definition**	**Cultivar Example** **Code** **Year**

1	Trumpet Daffodil cultivars	Solitary flower with corona as long as, or longer than the tepals	 'Little Gem' 1Y–Y 1959
2	Large-cupped Daffodil cultivars	Solitary flower with corona more than one-third, but less than equal to the length of the tepals	 'Fortune' 2Y–O 1923
3	Small-cupped Daffodil cultivars	Solitary flower with corona not more than one-third the length of the tepals	 'Barrett Browning' 3WWY–O 1945
4	Double Daffodil cultivars	One or more flowers with doubling of the corona or tepals or both	 'Butter and Eggs' 4Y–O 1777
5	Triandrus Daffodil cultivars	Two or more flowers with characteristics of *N. triandrus*, flowers pendent and tepals reflexed	 'Thalia' 5W–W 1916

6	Cyclamineus Daffodil cultivars	Solitary flower with characteristics of *N. cyclamineus*, flowers acutely angled to stem, very short pedicel, tepals reflexed	 'February Gold' 6Y–Y 1923
7	Jonquilla and Apodanthus Daffodil cultivars	One to five, but rarely up to eight flowers with characteristics of sections *Jonquilla* or *Apodanthi*, flowers acutely angled to stem, corona cup or funnel shaped or flared, and usually wider than long, tepals spreading or reflexed, usually fragrant	 'Stratosphere' 7Y–O 1968
8	Tazetta Daffodil cultivars	Three to twenty flowers, with characteristics of section *Tazetta*, flowers with tepals spreading, usually fragrant, stem stout	 'Scilly Valentine' 8Y–O 2000
9	Poeticus Daffodil cultivars	Usually solitary flower, with characteristics of *N. poeticus*, flowers with corona very short to disc shaped, usually green and or yellow centre and red rim (sometimes a single color), tepals white, usually fragrant	 'Actaea' 9W–YYR 1927

10	Bulbocodium Daffodil cultivars	Usually solitary flower, with characteristics of section *Bulbocodium*, flowers with tepals insignificant relative to corona, anthers dorsiflexed, filament and style usually curved	 'Golden Bells' 10Y–Y 1984
11	Split-corona Daffodil cultivars	Corona split, usually for more than half its length	
	(a) Collar Daffodils	Corona two whorls of three, opposite tepals	 'Mondragon' 11aY–O 1973
	(b) Papillon Daffodils	Corona usually single whorl of six, alternate to tepals	[Jodi] 'Jodi' 11bW–P/W 2002
12	Other Daffodil cultivars	Cultivars not fitting any of the above definitions	 'Tête-à-Tête' 12Y–Y 1949
13	Daffodils distinguished solely by botanical name	Wild (species) daffodils and hybrids without cultivar names	 *Narcissus × medioluteus* 1768

Garden Roses

Garden roses are predominantly hybrid roses that are grown as ornamental plants in private or public gardens. They are one of the most popular and widely cultivated groups of flowering plants, especially in temperate climates. Numerous cultivars have been produced, especially over the last two centuries, though roses have been known in the garden for millennia beforehand. While most garden roses are grown for their flowers, some are also valued for other reasons, such as having ornamental fruit, providing ground cover, or for hedging.

The Hybrid Tea rose, 'Peer Gynt'.

Cultivation

In the garden, roses are grown as bushes, shrubs or climbers. "Bushes" are usually comparatively low growing, often quite upright in habit, with multiple stems emerging near ground level; they are often grown formally in beds with other roses. "Shrubs" are usually larger and have a more informal or arching habit, and may additionally be placed in a mixed border or grown separately as specimens. Certain bush hybrids (and smaller shrubs) may also be grown as "standards", which are plants grafted high (typically 1 meter or more) on a rose rootstock, resulting in extra height which can make a dominant feature in a floral display. Climbing roses are usually trained to a suitable support.

Roses are commonly propagated by grafting onto a rootstock, which provides sturdiness and vigor, or (especially with Old Garden Roses) they may be propagated from hardwood cuttings and allowed to develop their own roots.

Most roses thrive in temperate climates. Those based on warm climate Asian species do well in their native sub-tropical environments. Certain species and cultivars can even flourish in tropical climates, especially when grafted onto appropriate rootstocks. Most garden roses prefer rich soil which is well-watered but well-drained, and perform best in well-lit positions which receive several hours of sun a day (although some climbers, some species and most Hybrid Musks will tolerate shade). Standard roses require staking.

Pruning

Rose pruning, sometimes regarded as a horticultural art form, is largely dependent on the type of rose to be pruned, the reason for pruning, and the time of year it is at the time of the desired pruning.

Most Old Garden Roses of strict European heritage (albas, damasks, gallicas, etc.) are shrubs that bloom once yearly, in late spring or early summer, on two-year-old (or older) canes. Their pruning requirements are quite minimal because removal of branches will remove next year's flower buds. Hence pruning is usually restricted to just removing weak and spent branches, plus light trimming (if necessary) to reduce overall size.

Modern hybrids, including the hybrid teas, floribundas, grandifloras, modern miniatures, and English roses, have a complex genetic background that almost always includes China roses (which are descended from *Rosa chinensis*). China roses were evergrowing, everblooming roses from humid subtropical regions that bloomed constantly on any new vegetative growth produced during the growing season. Their modern hybrid descendants exhibit similar habits; unlike Old European Roses, modern hybrids bloom continuously (until stopped by frost) on any new canes produced during the growing season. They therefore require pruning back of any spent flowering stem in order to divert the plant's energy into producing new growth and hence new flowers.

Additionally, Modern Hybrids planted in cold winter climates will almost universally require a "hard" annual pruning (reducing all canes to 8"–12", about 30 cm in height) in early spring. Again, because of their complex China rose background, modern hybrids are typically not as cold hardy as European Old Garden Roses, and low winter temperatures often desiccate or kill exposed canes. In spring, if left unpruned, these damaged canes will often die back all the way to the shrub's root zone, resulting in a weakened, disfigured plant. The annual "hard" pruning of hybrid teas and floribundas is generally done in early spring.

Deadheading

This is the practice of removing any spent, faded, withered, or discolored flowers. The purpose is to encourage the plant to focus its energy and resources on forming new shoots and blooms, rather than fruit production. Deadheading may also be performed for aesthetic purposes, if spent flowers are unsightly. Any roses such as Rosa glauca or Rosa moyesii that are grown for their decorative hips should not be deadheaded.

Pests and Diseases

Roses are subject to several diseases. The main fungal diseases affecting the leaves are rose black spot (*Diplocarpon rosae*), rose rust (*Phragmidium mucronatum*), rose powdery mildew (*Sphaerotheca pannosa*) and rose downy mildew (*Peronospora sparsa*). Stems can be affected by several canker diseases, the most commonly seen of which is stem canker (*Leptosphaeria coniothyrium*). Diseases of the root zone include honey fungus (*Armillaria* spp.), verticillium wilt, and various species of *phytophthora*.

Fungal leaf diseases affect some cultivars and species more than others. On susceptible plants fungicidal sprays may be necessary to prevent infection or reduce severity of attacks. Cultivation techniques may also be used, such as ensuring good air circulation around a plant. Stem cankers are best treated by pruning out infection as soon as it is noticed. Root diseases are not usually possible to treat once infection has occurred; the most practical line of defence is to ensure that growing conditions maximize plant health and thereby prevent infection. *Phytophthora* species are waterborne and therefore improving drainage and reducing waterlogging can help reduce infection.

The main pest affecting roses is the aphid (greenfly), which sucks the sap and weakens the plant. In areas where they are endemic Japanese beetles (*Popillia japonica*) take a heavy toll on rose flowers and foliage; rose blooms can also be destroyed by infestations of thrips (*Thysanoptera* spp). Roses are also used as food plants by the larvae of some Lepidoptera (butterfly and moth) species; see list of Lepidoptera that feed on roses. Spraying with insecticide of roses is often recommended but if this is done care is needed to minimize the loss of beneficial insects; systemic insecticides have the advantage of only affecting insects which feed on the plants.

Notable Rose Growers

Some rose growers are known for their particular contributions to the field. These include:

- David Austin nursery, based in Shropshire, UK, is the developer of "English roses", such as 'Constance Spry', 'Mary Rose' and 'Graham Thomas'
- Peter Beales was a specialist in classic and species roses, preserving many old and wild roses at his Norfolk nursery and also introducing 70 new cultivars. He was also the author of several classic books on the subject of roses.
- Joséphine de Beauharnais (Empress Josephine) was the first great collector of roses in the modern Western world, and her horticulturalist André Dupont pioneered the development of new hybrids using controlled pollination at her Malmaison estate. She has been called the godmother of modern rosomaniacs.
- Cants of Colchester, in Essex, is the UK's oldest firm of commercial rose growers. Notable introductions include 'Mrs B.R. Cant' and 'Just Joey'.
- Conard-Pyle Co. introduced the rose 'Peace' to the US and established the marque Star Roses. 'Peace' was bred by Meilland of France (where it was introduced as 'Mme A. Meilland'); Conard-Pyle acted as Meilland's US agents, and the rose was renamed for the US market when it was introduced at the end of the Second World War.
- Georges Delbard of Allier, France is more famous for new varieties of fruit tree, but among his nursery's roses are 'Centenaire de Lourdes', 'Altissimo' and 'Papa Delbard'.
- Dickson Roses, located near Belfast introduced its first roses in 1886, focusing on breeding Hybrid Teas that could stand up to the Irish climate. Successes include 'Shot Silk' and 'Grandpa Dickson' and, more recently, 'Elina' and 'Tequila Sunrise'.

- Pedro Dot put Spanish rose growing on the map and is best known for the shrub 'Nevada' and his work to improve the flower shape of miniature varieties.
- Claude Ducher was a Lyon hybridiser and nurseryman (father-in-law of Joseph Pernet-Ducher), whose roses include the Noisette 'Reve d'Or' and the Tea rose 'Marie van Houtte'.
- Andre Dupont was a French horticulturalist who pioneered the creation of new rose cultivars through controlled pollination. He was employed by the Empress Josephine to use her collection of roses to create new roses.
- Rudolf Geschwind was an Austro-Hungarian amateur rose breeder who introduced around 140 new varieties, including 'Gruss an Teplitz'. He focused on winter hardiness and vigour.
- Jules Gravereaux, founder of Roseraie de L'Haÿ
- Harkness Roses, in Hertfordshire, UK is best known for 'Ena Harkness' (at one time reputed to be the best-selling red Hybrid Tea in the world and actually bred by amateur rosarian Albert Norman). Other famous introductions include 'Compassion' and 'Margaret Merril'.
- S. Reynolds Hole was Dean of Rochester Cathedral in the UK and the founder of the (Royal) National Rose Society. He organized the first specialty rose show in the UK and published books on rose cultivation, popularizing rose growing and exhibiting.
- Jackson & Perkins is a hugely influential American rose grower. The company's early success was 'Dorothy Perkins', but under Eugene Boerner the focus on developing Floribundas led to many All-America Rose Selection honours.
- W. Kordes' Sons, based in Sparrieshoop in Schleswig-Holstein, Germany, is one of the most innovative rose breeders and growers, and responsible for the early flowering "Frühlings" series, the Kordesii Hybrids and many famous Hybrid Tea and Floribunda roses, including 'Crimson Glory' and 'Iceberg' ('Schneewittchen').
- McGredy, of Northern Ireland, was responsible for 'Evelyn Fison', 'Dublin Bay' and also 'Regensberg', a pioneering 'handpainted' rose. Sam McGredy IV moved to New Zealand in 1974 and focused on hybrid teas and Grandifloras, including 'Paddy Stephens' and 'Kathryn McGredy'.
- Meilland family made its name and fortune with 'Mme A. Meilland' ('Peace'), and has continued to be at the forefront of rose breeding, with varieties such as 'Bonica '82' and 'Swany'.
- Ralph S. Moore, the California-based breeder of more than 500 roses, is known as 'the father of Modern Miniatures' and was a hugely influential figure in the development of commercial approaches to rose hybridization.
- Werner Noack of Germany introduced the Flower Carpet (ground cover/landscape) series.
- Paul was a Hertfordshire, UK nursery that involved two brothers (William and George) and their two sons (Arthur William and Paul Laing). Collectively, Paul is known today for varieties such as 'Paul's Lemon Pillar' and 'Paul's Scarlet

Climber'. Experimental hybrids using species roses resulted in choice varieties such as 'Mermaid'.

- Joseph Pemberton was an Anglican clergyman and amateur rosarian who set out to breed 'old fashioned' roses. The resulting hybrid musks include 'Felicia' and 'Penelope'. On his death, the nursery passed to his gardener J.A. Bentall, who produced 'Buff Beauty' and the Polyantha 'The Fairy'.
- Jean Pernet, père was a Lyon nurseryman whose notable roses include the Moss variety 'Louis Gimard' and the hybrid perpetual 'Baronne Adolphe de Rothschild'.
- Joseph Pernet-Ducher was among the first rose breeders to focus on developing the new Hybrid Tea class. His introductions include 'Mme Caroline Testout' and 'Soleil d'Or'- forerunner of 20th-century yellow and orange roses.
- Poulson, the Danish rose dynasty, was established in 1878 and originally focused on breeding roses hardy enough to withstand the Scandinavian climate. Later introductions notable for their form and color include 'Chinatown' (1963) and 'Ingrid Bergman' (1984). The nursery developed a number of successful ground cover (landscape) roses, including 'Kent' (1988).
- Rose Barni in Tuscany specialises in roses for Mediterranean climates. Notable successes include 'Castore' and 'Polluce', and striped varieties such as 'Rinascimento' and 'Missoni'.
- Mathias Tantau
- Graham Stuart Thomas is best known for reawakening interest in old garden roses, but also ensured commercial introductions in the wild rose style, including 'Bobbie James' and 'Souvenir de St Anne's'.
- Dr. Walter van Fleet worked for the US Department of Agriculture, focusing on crops, but also developing roses designed to thrive in the American climate. His introductions include 'American Pillar' and 'Dr W. Van Fleet'. After his death, his seedlings - including 'Mary Wallace', 'Breeze Hill' and 'Glenn Dale' - were introduced by the American Rose Society as 'dooryard climbers'.
- Jean-Pierre Vibert was a prolific early rose hybridizer, responsible for many older roses still found in gardens today. 'Aimee Vibert' (1828), one of his Noisettes, was named for his daughter.
- Weeks Roses (with Tom Carruth) is a California rose company that has focused on innovations in color, form and vigour. Its roses include 'Night Time', 'Stainless Steel', 'Fourth of July' and 'Hot Cocoa'.

Tree Breeding

Tree breeding is the application of genetic, reproductive biology and economics principles to the genetic improvement and management of forest trees. In contrast to the selective breeding of livestock, arable crops, and horticultural flowers over the last few centuries, the breeding of trees, with the exception of fruit trees, is a relatively recent occurrence.

A typical forest tree breeding program starts with selection of superior phenotypes (plus trees) in a natural or planted forest. This application of mass selection improves the mean performance of the forest. Offspring is obtained from selected trees and grown in test plantations that act as genetic trials. Based on such tests the best genotypes among the parents can be selected. Selected trees are typically multiplied by either seeds or grafting and seed orchards are established when the preferred output is improved seed. Alternatively, the best genotypes can be directly propagated by cuttings or in-vitro methods and used directly in clonal plantations. The first system is frequently used in pines and other conifers, while the second is typical in some broadleaves (poplars, eucalypts and others). The objectives of a tree breeding program range from yield improvement and adaptation to particular conditions, to pest- and disease-resistance, wood properties, etc. Currently, tree breeding starting to take advantage of the fast development in plant genetics and genomics.

Optimization

Tree breeders make efforts to get their operation efficient by optimizing tree breeding. Scientists develop tools aimed at improvement of the efficiency of tree breeding programmes. Optimizing can mean adapting strategies and methods to certain species, group of populations, structure of genetic variation and mode of inheritance of the important traits to obtain the highest benefit per unit of time. Optimizing is usually carried out at the following levels: breeding strategy (appropriate intensity of breeding, breeding population structure and size, plan for maintenance of genetic diversity), breeding methods (mating type, testing and selection methods, testing population size and time) and deployment methods of the genetically improved material (seed orchards and clonal forestry: genetic contribution, size). Computer simulators are frequently used: stochastic – based on defined and random algorithms; and deterministic – based on defined algorithms.

Selection strategies have been compared for annual progress in long-term breeding at a given annual cost considering genetic gain, gene diversity, cost components, and time components. For Norway spruce it seems favorable to clone full sib families and then select based on clonal performance while for Scots pine a two-stage strategy seems best, first phenotypic pre-selection and then progeny-testing the selections.

Tree Improvement

A genetically variable population and a method of selecting genetically superior individuals provide the basis for tree improvement by breeding. In essence, a tree improvement program sets out to isolate and evaluate the genetic component of variation in one or more characters of interest. In the simplest procedure, cycles of selection reduce the available population in a particular direction to enhance desirable traits, then breeding from selections to expand the population with improved characteristics. Breeding strategies vary with species and objectives, but all use mating designs to generate information and new material. Choice of a suitable breeding strategy and mating design is a key decision in any breeding program. Kiss (1986)used a 2-level design in British Columbia to study variation within and between separate populations of white spruce, both within British Columbia and from eastern North America.

The breeding program for white spruce initiated in 1986 by the Canadian Forestry Service in the Maritimes employed 2 kinds of mating: polycross, to test clones for general combining ability; and pair-mating, to generate material for second generation selections (Fowler et al. 1988).

Newton's (2003) systematic review of yield responses of white spruce and 3 other North American conifers to forest tree improvement practices indicated that correct provenance-progeny selection could yield juvenile height growth gains of about 12% at 20 years for white spruce, and a corresponding merchantable productivity (mean annual merchantable volume increment) gain of 26% at 50 years for plantations established at nominal initial densities on medium-to-good quality sites. Also, preliminary estimates derived from individual case studies indicated that first generational selection strategies for white spruce could increase merchantable productivity by approximately 20% at 45 years.

CULTIGEN

A cultigen (from the Latin *cultus* – cultivated, and *gens* – kind) is a plant that has been deliberately altered or selected by humans; it is the result of artificial selection. These "man-made" or anthropogenic plants are, for the most part, plants of commerce that are used in horticulture, agriculture and forestry. Because cultigens are defined by their mode of origin and not by where they are growing, plants meeting this definition remain cultigens whether they are naturalised in the wild, deliberately planted in the wild, or growing in cultivation. Cultigens arise in the following ways: selections of variants from the wild or cultivation including vegetative sports (aberrant growth that can be reproduced reliably in cultivation); plants that are the result of plant breeding and selection programs; genetically modified plants (plants modified by the deliberate implantation of genetic material); and graft-chimaeras (plants grafted to produce mixed tissue, the graft material possibly from wild plants, special selections, or hybrids).

Cultigens may be named in any of a number of ways. The traditional method of scientific naming is under the *International Code of Nomenclature for algae, fungi, and plants*, and many of the most important cultigens, like maize (*Zea mays*) and banana (*Musa acuminata*), are so named. Although it is perfectly in order to give a cultigen a botanical name, in any rank desired, now or at any other time, these days it is more common for cultigens to be given names in accordance with the principles, rules and recommendations laid down in the International Code of Nomenclature for Cultivated Plants (ICNCP) which provides for the names of cultigens in three classification categories, the cultivar, the Group (formerly cultivar-group), and the grex. From that viewpoint it may be said that there is a separate discipline of cultivated plant taxonomy, which forms one of the ways to look at cultigens. The ICNCP does not recognize the use of trade designations and other marketing devices as scientifically acceptable names, but does provide advice on how they should be presented.

Not all cultigens have been given names according to the *Cultivated Plant Code*. Apart from ancient cultigens like those mentioned above there may be occasional anthropogenic plants such as those that are the result of breeding, selection, and tissue grafting that are of no commercial value and have therefore not been given names according to the ICNCP.

Formal Definition

A cultigen is a plant whose origin or selection is primarily due to intentional human activity.

The Wild/Cultivated Distinction

Interest in the distinction between wild and cultivated plants dates back to antiquity. Botanical historian Alan Morton notes that wild and cultivated plants (cultigens) were of intense interest to the ancient Greek botanists (partly for religious reasons) and that the distinction was discussed in some detail by Theophrastus (370–285 BCE) the "Father of Botany". Theophrastus was a pupil of both Plato and Aristotle and succeeded the latter as head of the Peripatetic School of Philosophy at the Lyceum in Athens. Theophrastus accepted the view that it was human action, not divine intervention, that produced cultivated plants (cultigens) from wild plants and he also "*had an inkling of the limits of culturally induced (phenotypic) changes and of the importance of genetic constitution*" (*Historia Plantarum* III, 2,2 and *Causa Plantarum* I, 9,3). He also noted that cultivated varieties of fruit trees would degenerate if cultivated from seed.

Origin of Term

The word cultigen was coined in 1918 by Liberty Hyde Bailey (1858–1954) an American horticulturist, botanist and cofounder of the American Society for Horticultural Science. He was aware of the need for special categories for those cultivated plants that had arisen by intentional human activity and which would not fit neatly into the Linnaean hierarchical classification of ranks used by the *International Rules of Botanical Nomenclature* (which later became the *International Code of Nomenclature for algae, fungi, and plants*).

In his 1918 paper Bailey noted that for anyone preparing a descriptive account of the cultivated plants of a country (he was at that time preparing such an account for North America) it would be clear that there are two *gentes* or kinds (Latin singular, *gens*; plural, *gentes*) of plants. Firstly, those that are of known origin or nativity "of known habitat". These he referred to as indigens. The other kind was:

" ... a domesticated group of which the origin may be unknown or indefinite, which has such characters as to separate it from known indigens, and which is probably not represented by any type specimen or exact description, having therefore no clear taxonomic beginning."

He called this second kind of plant a cultigen, the word derived from the conflation of the Latin *cultus* – cultivated, and *gens* – kind.

In 1923 Bailey extended his original discussion emphasising that he was dealing with plants at the rank of species and he referred to indigens as:

"*those that are discovered in the wild*"

and cultigens as plants that:

"*arise in some way under the hand of man*"

He then defined a cultigen as:

"a species, or its equivalent, that has appeared under domestication"

Bailey's Definitions

Bailey soon altered his 1923 definition of cultigen when, in 1924, he gave a new definition in the Glossary of his *Manual of Cultivated Plants* as:

" *Plant or group known only in cultivation; presumably originating under domestication; contrast with indigen* "

This, in essence, is the definition given at the head of this piece. This definition of the cultigen permits the recognition of cultivars, unlike the 1923 definition which restricts the idea of the cultigen to plants at the rank of species.

In later publications of the *Liberty Hyde Bailey Hortorium*, Cornell, the idea of the cultigen having the rank of species returned (e.g. *Hortus Second* in 1941 and *Hortus Third* in 1976): both of these publications indicate that the terms cultigen and cultivar are not synonymous and that cultigens exist at the rank of species only.

"*A cultigen is a plant or group of apparent specific rank, known only in cultivation, with no determined nativity, presumably having originated, in the form in which we know it, under domestication. Compare indigen. Examples are* Cucurbita maxima, Phaseolus vulgaris, Zea mays ".

Recent usage in horticulture has, however, maintained a distinction between cultigen and cultivar while nevertheless *allowing the inclusion of cultivars* within the definition.

Cultivars

Cultigen and cultivar may be confused with one another. Cultigen is a general-purpose term encompassing not only plants with cultivar names but others as well, while cultivar is a formal classification category (in the ICNCP).

Although in his 1923 paper Bailey used only the rank of species for the cultigen, it was clear to him that many domesticated plants were more like botanical varieties than species and so he established a new classification category for these, the cultivar, generally assumed to be a contraction of the words "cultivated" and "variety". Bailey was never explicit about the etymology of the word cultivar and it has been suggested that it is a contraction of the words "cultigen" and "variety". He defined cultivar in his 1923 paper as:

... "*a race subordinate to species, that has originated and persisted under cultivation; it is not necessarily, however, referable to a recognised botanical species. It is essentially the equivalent of the botanical variety except in respect to its origin*".

This definition and understanding of cultivar has changed over time.

Usage

Usage in Botany

In botanical literature the word cultigen is generally used to denote a plant which, like the bread wheat (*Triticum aestivum*) is of unknown origin, but presumed to be an ancient human selection. Plants like bread wheat have been given binomials according to the *Botanical Code* and therefore have names with the same form as those of plant species that occur naturally in the wild, but it is not necessary for a cultigen to have a species name, or to have the biological characteristics that distinguish a species. Cultigens can have names at any of various other ranks, including cultivar names, names in the classification categories of grex and group, variety names, forma names, or they may be plants that have been altered by humans (including genetically modified plants) but which have not been given formal names.

Usage in Horticulture

The year 1953 was an important one for cultivated plant taxonomy because this was the date of publication of the first *International Code of Nomenclature for Cultivated Plants* in which Bailey's term cultivar was introduced. It was also the year that the eponymous journal commemorating the work of Bailey (who died in 1954), *Baileya*, was published. In the first volume of *Baileya* taxonomist and colleague of Bailey, George Lawrence, wrote a short article clarifying the distinction between the new term *cultivar* and the *variety*. In the same article he also tried to clarify the critical term *taxon* which had been introduced by German biologist Meyer in the 1920s but had only just been introduced and accepted in botanical circles. This brief article by Lawrence is useful for its insight into the understanding of the meaning of the word cultigen at this time. He opens the article:

In 1918, L.H. Bailey distinguished those plants originating in cultivation from the native plants by designating the former as cultigens and the latter as indigens (indigenous or native to the region). At the same time he proposed the term cultivar to distinguish varieties originating in cultivation from botanical varieties known first in the wild.

In horticulture the definition and use of the terms cultigen and cultivar has varied. One example is the definition given in the Botanical Glossary of *The New Royal Horticultural Society Dictionary of Gardening* which defines cultigen as:

" *A plant found only in cultivation or in the wild having escaped from cultivation; included here are many hybrids and cultivars,* "...

The use of cultigen in this sense is essentially the same as the definition of the cultigen published by Bailey in 1924.

The *Cultivated Plant Code*, however, states that cultigens are "maintained as recognisable entities solely by continued propagation", and thus would not include plants that have evolved subsequent to escape from cultivation.

Recommended Usage

Wider use of the term cultigen as defined here has been proposed for the following reasons:

- supports current usage in horticulture
- assists clarity in non-technical discussions about "wild" and "cultivated" plants (for example, cultivated plants as commonly understood (plants in cultivation) are not the same as the "cultivated plants" of the *Cultivated Plant Code*, and the distinction between "wild" and "cultivated" habitats is becoming progressively blurred)
- has the potential to simplify the language and definitions used in the Articles and Recommendations of the *Cultivated Plant Code*
- gives greater precision and clarity to the definition of the respective scope, terminology and concepts of the *Botanical Code* and the *Cultivated Plant Code*
- avoids the potential for confusion within the *Cultivated Plant Code* over its scope, that is, whether it is concerned with:
- where plants are growing (in the wild or in cultivation)
- how they originated (whether they are the result of intentional human activity or not)
- whether it simply provides a mechanism for regulating the names of those cultigens requiring special classification categories that are not part of the Linnaean hierarchy of the *Botanical Code* i.e. cultivar and Group names

Critique of Definition

Potential misunderstandings and questions arising from the definition of cultigen given here have been discussed in the literature and are summarized below.

- Natural and *artificial* selection

The selection process is termed "artificial" when human preferences or influences have a significant effect on the evolution of a particular population or species. Note: artificial selection is a part of the overall selection process – it does not imply that humans are not part of nature, it is simply useful sometimes to distinguish when there has been human influence on selection (as with cultigens).

- What exactly does *altered* mean?

There are cases that do not seem to comply with the definition. For example, we can presume that the entire global flora is changing as a result of human-induced climate change. Does this mean that all plants are cultigens?

In cases like this the definition refers to "deliberate" selection and this would be of particular plant characteristics that are not exhibited by a plant's wild counterparts.

- What exactly does *deliberately selected* mean?

From the moment a plant is taken from the wild it is subject to human selection pressure – from the selection of the original propagation material to the purchase of the plant in a nursery. Surely this form of selection is not deliberate? Again, the early human selection of crops 7,000-10,000 years ago is thought to have occurred quite unintentionally. Variants useful to horticulture often arise spontaneously, they are not deliberate products. Are these cases of unintentional, accidental, or unconscious selection?

There certainly appear to be cases where origin or selection of a plant is not "deliberate". However, the long term propagation of plants that have some utility, usually economic or ornamental, can hardly be regarded as unintentional and these plants will, almost without exception, have characteristic(s) that distinguish them from their wild counterparts.

- What about plants selected from the wild?

Plants like *Quercus robur*, Pedunculate or English Oak, *Liquidambar styraciflua*, Sweetgum and *Eucalyptus globulus*, Blue Gum grown in parks and gardens are essentially the same as their wild counterparts and are therefore not cultigens. However, occasionally within natural plant variation there occur characters that are of value to horticulture but of little interest to botany. For example a plant might have flowers of several different colors but these may not have been given formal botanical names. It is customary in horticulture to introduce such variants to commerce and to give them cultivar names. Technically these plants have not been deliberately altered in any way from plants growing (or once growing) in the wild but as they are *deliberately selected* and *named* it seems permissible to refer to them as cultigens. These occurrences are very few. The definition could be (clumsily) extended by mentioning that selection can be for "desirable variation that is not recognized in botanical nomenclature".

- What about gene flow between populations?

Occasionally cultigens escape from cultivation into the wild where they breed with indigenous plants. Selections may be made from the progeny in the wild and brought back into cultivation where they are used for breeding and the results of the breeding again escape into the wild to breed with indigenous plants. *Lantana* has behaved much like this. The genetic material of a cultigen may become part of the gene pool of a population where, over time, it may be largely or completely swamped. In cases like this what plants are to be called cultigens?

Whether a plant is a cultigen or not does not depend on where it is growing. If it complies with the definition then it is a cultigen. Cases like this have always been difficult for botanical nomenclature. Unnamed progeny in the wild might be given a name like *Lantana* aff. *camara* (aff. = having affinities with) or may remain unnamed. Its cultigenic origin may or may not be recognized by the allocation of a cultivar name.

- Plants of unknown origin

Occasionally plants will occur whose origin is unknown. Plants growing in cultivation that are unknown in the wild may be determined as cultigenic as a result of scientific investigation, but may remain a mystery.

- Difficult cases

It may happen that a hybrid cross that has occurred in nature is also performed deliberately in cultivation and that the progeny appear identical. How do we know which plants are cultigens?

If the cross in cultivation is followed by deliberate selection and naming then this will indicate a cultigen. However in a case like this it may not be possible to tell.

REFERENCES

1. Bailey 1923, p. 113
2. Spencer & Cross 2007, p. 938
3. Lawrence 1953, pp. 19–20
4. Cultivated Plant Code Article 14.1 Brickell 2009, p. 19
5. Morton 1981, pp. 38–39
6. Lawrence 1955, p. 177
7. Lawrence 1955, p. 180
8. Lawrence 1955, p. 181
9. Lawrence 1955, pp. 179–180
10. Bailey 1923, p. 113
11. Trehane 2004, p. 17
12. Lawrence 1960, p. 1
13. Cultivated Plant Code Art. 2.3 Brickell 2009, p. 1
14. Spencer, Cross & Lumley 2007, p. 47
15. Cultivated Plant Code. Art. 2.1 Brickell 2009, p. 6
16. Cultivated Plant Code. Art. 4 Brickell 2009, p. 12
17. Cultivated Plant Code. Art. 3 Brickell 2009, pp. 10–12
18. Cultivated Plant Code. Art. 2.2 Brickell 2009, p. 6
19. Cultivated Plant Code. Preamble & Principles Brickell 2009, p. 19
20. Cultivated Plant Code, Article 2.20 Brickell 2009, p. 9
21. Cultivated Plant Code, Articles 2.5–2.11 Brickell 2009, pp. 6–7
22. Courses / RHS Gardening
23. Cultivated Plant Code, Articles 2.17–2.18 Brickell 2009, pp. 7–8
24. Cultivated Plant Code, Articles 2.12–2.16 Brickell 2009, pp. 7–8
25. Cultivated Plant Code, Articles 2.19 Brickell 2009, pp. 8–9
26. Cultivated Plant Code Recommendation 8A.1 Brickell 2009, p. 15
27. Cultivated Plant Code Article 21.3 Brickell 2009, p. 25
28. Cultivated Plant Code Art. 14 Brickell 2009, p. 19
29. Cultivated Plant Code Art. 3 Brickell 2009, pp. 10–12
30. Cultivated Plant Code Art. 15 Brickell 2009, p. 19
31. P. Gepts (2004) Who Owns Biodiversity, and How Should the Owners Be Compensated? Plant Physiology 134, pp. 1295–1307

32. BSPB Plant breeding – The business and science of crop improvement British Society of Plant Breeders booklet
33. Adi, A.B.C., Intellectual Property Rights in Biotechnology and the Fate of Poor Farmers' Agriculture. Social Science Research Network
34. 'Who owns nature?' (article by nurseryman and plant hunter Michael Wickenden, published in The Plantsman)
35. Spencer, Cross & Lumley 2007, pp. 76–81
36. Spencer, Cross & Lumley 2007, p. 78
37. Cultivated Plant Code Brickell 2009, pp. 62, 67–83
38. Staff (2010). "ISHS :: Commission Nomenclature and Cultivar Registration - International Cultivar Registration Authorities (ICRAs)". ishs.org. Retrieved 5 March 2011.
39. Bailey, Liberty Hyde (1923). "Various cultigens, and transfers in nomenclature". Gentes Herbarum. 1 (Part 3): 113–136.
40. Brickell, Chris D. et al. (eds) (2009). "International Code of Nomenclature for Cultivated Plants (ICNCP or Cultivated Plant Code) incorporating the Rules and Recommendations for naming plants in cultivation. 8th ed., adopted by the International Union of Biological Sciences International Commission for the Nomenclature of Cultivated Plants" (PDF). Scripta Horticulturae. International Society of Horticultural Science. 10: 1–184. ISBN 978-90-6605-662-6.
41. Lawrence, George H.M. (1953). "Cultivar, Distinguished from Variety". Baileya. 1: 19–20.
42. Lawrence, George H.M. (1955). "The Term and Category of Cultivar". Baileya. 3: 177–181.
43. Lawrence, George H.M. (1957). "The Designation of Cultivar-names". Baileya. 5: 162–165.
44. Lawrence, George H.M. (1960). "Notes on Cultivar Names". Baileya. 8: 1–4.
45. Morton, Alan G. (1981). History of Botanical Science: An Account of the Development of Botany from Ancient Times to the Present Day. London: Academic Press. ISBN 0-12-508382-3.
46. Spencer, Roger; Cross, Robert; Lumley, Peter (2007). Plant names: a guide to botanical nomenclature. (3rd ed.). Collingwood, Australia: CSIRO Publishing (also Earthscan, UK.). ISBN 978-0-643-09440-6.
47. Spencer, Roger D.; Cross, Robert G. (2007). "The International Code of Botanical Nomenclature (ICBN), the International Code of Nomenclature for Cultivated Plants (ICNCP), and the cultigen". Taxon. 56 (3): 938–940. doi:10.2307/25065875.
48. Trehane, Piers (2004). "50 years of the International Code of Nomenclature for Cultivated Plants". Acta Horticulturae. 634: 17–27.
49. McNeill, J.; Barrie, F.R.; Buck, W.R.; Demoulin, V.; Greuter, W.; Hawksworth, D.L.; Herendeen, P.S.; Knapp, S.; Marhold, K.; Prado, J.; Prud'homme Van Reine, W.F.; Smith, G.F.; Wiersema, J.H.; Turland, N.J. (2012). International Code of

Nomenclature for algae, fungi, and plants (Melbourne Code) adopted by the Eighteenth International Botanical Congress Melbourne, Australia, July 2011. Regnum Vegetabile 154. A.R.G. Gantner Verlag KG. ISBN 978-3-87429-425-6. Art. 28 "Note 3. Nothing precludes the use, for cultivated plants, of names published in accordance with the requirements of this Code."

50. Brickell CD, Alexander C, David JC, Hetterscheid WA, Leslie AC, Malecot V, Jin X (2009). International Code of Nomenclature for Cultivated Plants (ICNCP or Cultivated Plant Code) incorporating the Rules and Recommendations for naming plants in cultivation, Eighth Edition, Adopted by the International Union of Biological Sciences International Commission for the Nomenclature of Cultivated Plants (PDF). Editorial committee; Cubey, J.J. International Association for Plant Taxonomy and International Society for Horticultural Science. Article 10 and Appendix 10.
51. Morton, A.G. 1981. History of Botanical Science. London: Academic Press. ISBN 0-12-508382-3.
52. Bailey, L.H. 1918. The indigen and cultigen. Science ser. 2, 47:306-308
53. Bailey, L.H. 1923. Various cultigens, and transfers in nomenclature. Gentes Herb. 1:113--136
54. Bailey, L.H. 1924. Manual of cultivated plants. Macmillan, New York
55. Bailey, L.H. 1941. Hortus second : a concise dictionary of gardening, general horticulture and cultivated plants in North America. The Macmillan Company, New York.
56. Bailey, L.H. 1976. Hortus third : a concise dictionary of plants cultivated in the United States and Canada. Macmillan Collier Macmillan, New York.
57. Trehane, P. 2004. 50 years of the International Code of Nomenclature for Cultivated Plants. Acta Horticulturae 634: 17-27.
58. Spencer, Cross & Lumley 2007, p. 47
59. Lawrence, George H. M. 1953. "Cultivar, Distinguished from Variety." Baileya 1: 19–20.
60. Huxley, A., Griffiths, M., and Levy, M. (eds.) 1992. The New Royal Horticultural Society Dictionary of Gardening. Macmillan, London
61. Spencer, R.D. 1999. Cultivated plants and the codes of nomenclature – towards the resolution of a demarcation dispute. pp.171—181 in: S. Andrews, A.C. Leslie and C. Alexander (eds). Taxonomy of Cultivated Plants: Third Symposium. Royal Botanic Gardens, Kew. ISBN 1-900347-89-X

CHAPTER 3

Molecular Breeding

Chapter Contents

- **Introduction**
- **Quantitative Trait Locus**
- **Marker-Assisted Selection**
- **Genetic Engineering**

Molecular breeding (MB) may be defined in a broad-sense as the use of genetic manipulation performed at DNA molecular levels to improve characters of interest in plants and animals, including genetic engineering or gene manipulation, molecular marker-assisted selection, genomic selection, etc. More often, however, molecular breeding implies molecular marker-assisted breeding (MAB) and is defined as the application of molecular biotechnologies, specifically molecular markers, in combination with linkage maps and genomics, to alter and improve plant or animal traits on the basis of genotypic assays. This term is used to describe several modern breeding strategies, including marker-assisted selection (MAS), marker-assisted backcrossing (MABC), marker-assisted recurrent selection (MARS), and genome-wide selection (GWS) or genomic selection (GS)

INTRODUCTION

Molecular breeding is the application of molecular biology tools, often in plant breeding.

The areas of molecular breeding include:

- QTL mapping or gene discovery

- Marker assisted selection and genomic selection
- Genetic engineering
- Genetic transformation

STEPS

Marker Assisted Breeding

Genotyping and creating molecular maps - genomics

The commonly used markers include Simple sequence repeats (or microsatellites), single nucleotide polymorphisms (SNP). The process of identification of plant genotypes is known as genotyping.

Development of SNPs has revolutionized the molecular breeding process as it help to create dense markers. Another area that is developing is genotyping by sequencing.

Phenotyping - phenomics

To identify genes associated with traits, it is important to measure trait value - known as phenotype. "omics" for measurement of phenotypes is called phenomics. The phenotype can be indicative of measurement of trait itself or an indirectly related or correlated trait.

QTL Mapping or Association Mapping

Genes (Quantitative trait loci (abbreviated as QTLs) or quantitative trait genes or minor genes or major genes) involved in controlling trait of interest is identified. The process is known as mapping. Mapping of such genes can be done using molecular markers. QTL mapping can involve single large family, unrelated individuals or multiple families (see: Family based QTL mapping). Basic idea is to identify genes or markers associated with genes that can be used in marker assisted breeding / selection

Marker Assisted Selection or Genetic Selection

Once genes or markers are identified they can be used for genotyping and selection decision can be made.

Marker-Assisted Backcrossing (MABC)

Backcross is crossing F1 with its parents to transfer a limited number of loci (e.g. transgene, disease resistance loci, etc.) from one genetic background to another. Usually the recipient of such genes is good adopted cultivars otherwise except the gene is to be transferred. So we want to keep genetic background of the recipient genotypes, which is done by 4-6 rounds of repeated backcrosses while selecting for the gene of interest. We can use markers from the whole genome to recover the genome quickly in 2-3 rounds of backcrossing might be good enough in such situation.

Marker-Assisted Recurrent Selection (MARS)

MARS include identification and selection of several genomic regions (up to 20 or even more) for complex traits within a single population.

Genomic selection

Genomic selection is novel approach to traditional marker-assisted selection where selection are made based on few markers. Rather than seeking to identify individual loci significantly associated with a trait, genomics uses all marker data as predictors of performance and consequently delivers more accurate predictions. Selection can be based on genomic selection predictions, potentially leading to more rapid and lower cost gains from breeding.Genomic prediction combines marker data with phenotypic and pedigree data (when available) in an attempt to increase the accuracy of the prediction of breeding and genotypic values.

Genetic Transformation or Genetic Engineering

Transfer of genes make it possible for horizontal transfer of genes from one organism to another. Thus plants can receive genes from humans or algae or any other organism. This provides limitless opportunity in breeding crop plants.

QUANTITATIVE TRAIT LOCUS

A quantitative trait locus (QTL) is a section of DNA (the locus) that correlates with variation in a phenotype (the quantitative trait). Usually the QTL is linked to, or contains, the genes that control that phenotype. QTLs are mapped by identifying which molecular markers (such as SNPs or AFLPs) correlate with an observed trait. This is often an early step in identifying and sequencing the actual genes that cause the trait variation.

Quantitative traits are phenotypes (characteristics) that vary in degree and can be attributed to polygenic effects, i.e., the product of two or more genes, and their environment.

History

Mendelian inheritance was rediscovered at the beginning of the 20th century, and as Mendel's ideas spread geneticists began to connect Mendel's rules of inheritance of single factors to Darwinian evolution. For early geneticists, it was not immediately clear that the smooth variation in traits like body size (i.e., Incomplete Dominance) was caused by the inheritance of single genetic factors. Although Darwin himself observed that inbred features of fancy pigeons were inherited in accordance with Mendel's laws (although Darwin didn't actually know about Mendel's ideas when he made the observation), it was not obvious that these features selected by fancy pigeon breeders can similarly explain quantitative variation in nature.

An early attempt by William Ernest Castle to unify the laws of Mendelian inheritance with Darwin's theory of speciation invoked the idea that species become distinct from one another as one species or the other acquires a novel Mendelian factor. Castle's conclusion was based on the observation that novel traits that could be studied in the lab and that show Mendelian inheritance patterns reflect a large deviation from the wild type, and Castle believed that acquisition of such features is the basis of "discontinuous variation" that characterizes speciation. Interestingly, Darwin discussed the inheritance of similar mutant features but did not invoke them as a requirement of speciation. Instead Darwin used the emergence of such features in breeding populations as evidence that mutation can occur at random within breeding populations, which is a central premise of his model of selection in nature. Later in his career, Castle would refine his model for speciation to allow for small variation to contribute to speciation over time. He also was able to demonstrate this point by selectively breeding laboratory populations of rats to obtain a hooded phenotype over several generations.

Castle's was perhaps the first attempt made in the scientific literature to direct evolution by artificial selection of a trait with continuous underlying variation, however the practice had previously been widely employed in the development of agriculture to obtain livestock or plants with favorable features from populations that show quantitative variation in traits like body size or grain yield.

Castle's work was among the first to attempt to unify the recently rediscovered laws of Mendelian inheritance with Darwin's theory of evolution. Still, it would be almost thirty years until the theoretical framework for evolution of complex traits would be widely formalized. In an early summary of the theory of evolution of continuous variation, Sewall Wright, a graduate student who trained under Castle, summarized contemporary thinking about the genetic basis of quantitative natural variation: "As genetic studies continued, ever smaller differences were found to mendelize, and any character, sufficiently investigated, turned out to be affected by many factors." Wright and others formalized population genetics theory that had been worked out over the preceding 30 years explaining how such traits can be inherited and create stably breeding populations with unique characteristics. Quantitative trait genetics today leverages Wright's observations about the statistical relationship between genotype and phenotype in families and populations to understand how certain genetic features can affect variation in natural and derived populations.

Quantitative Traits

Polygenic inheritance refers to inheritance of a phenotypic characteristic (trait) that is attributable to two or more genes and can be measured quantitatively. Multifactorial inheritance refers to polygenic inheritance that also includes interactions with the environment. Unlike monogenic traits, polygenic traits do not follow patterns of Mendelian inheritance (discrete categories). Instead, their phenotypes typically vary along a continuous gradient depicted by a bell curve.

An example of a polygenic trait is human skin color variation. Several genes factor into determining a person's natural skin color, so modifying only one of those genes can change skin color slightly or in some cases, such as for SLC24A5, moderately. Many disorders with genetic components are polygenic, including autism, cancer, diabetes and numerous others. Most phenotypic characteristics are the result of the interaction of multiple genes.

Examples of disease processes generally considered to be results of *multifactorial etiology*:

Congenital malformation

- Cleft palate
- Congenital dislocation of the hip
- Congenital heart defects
- Neural tube defects
- Pyloric stenosis
- Talipes

Adult onset diseases

- Diabetes Mellitus
- Cancer
- Glaucoma
- Hypertension
- Ischaemic heart disease
- Manic depression
- Schizophrenia
- Psoriasis
- Thyroid diseases
- Alzheimer's Disease

Multifactorially inherited diseases are said to constitute the majority of genetic disorders affecting humans which will result in hospitalization or special care of some kind.

Multifactorial Traits in General

Traits controlled by the both environment and genetic factors. Usually, multifactorial traits outside of illness result in what we see as continuous characteristics in organisms, especially human organisms such as: height,skin color, and body mass. All of these phenotypes are complicated by a great deal of give-and-take between genes and environmental effects. The continuous distribution of traits such as height and skin color described above, reflects the action of genes that do not manifest typical patterns of dominance and recessiveness. Instead the contributions of each involved locus are thought to be additive. Writers have distinguished this kind of inheritance as *polygenic*, or *quantitative inheritance*.

Thus, due to the nature of polygenic traits, inheritance will not follow the same pattern as a simple monohybrid or dihybrid cross. Polygenic inheritance can be explained as Mendelian inheritance at many loci, resulting in a trait which is normally-distributed. If

n is the number of involved loci, then the coefficients of the binomial expansion of $(a + b)^{2n}$ will give the frequency of distribution of all n allele combinations. For a sufficiently high n, this binomial distribution will begin to resemble a normal distribution. From this viewpoint, a disease state will become apparent at one of the tails of the distribution, past some threshold value. Disease states of increasing severity will be expected the further one goes past the threshold and away from the mean.

Heritable Disease and Multifactorial Inheritance

A mutation resulting in a disease state is often recessive, so both alleles must be mutant in order for the disease to be expressed phenotypically. A disease or syndrome may also be the result of the expression of mutant alleles at more than one locus. When more than one gene is involved, with or without the presence of environmental triggers, we say that the disease is the result of multifactorial inheritance. The more genes involved in the cross, the more the distribution of the genotypes will resemble a normal, or Gaussian distribution. This shows that multifactorial inheritance is polygenic, and genetic frequencies can be predicted by way of a polyhybrid Mendelian cross. Phenotypic frequencies are a different matter, especially if they are complicated by environmental factors.

The paradigm of polygenic inheritance as being used to define multifactorial disease has encountered much disagreement. Turnpenny (2004) discusses how simple polygenic inheritance cannot explain some diseases such as the onset of Type I diabetes mellitus, and that in cases such as these, not all genes are thought to make an equal contribution.

The assumption of polygenic inheritance is that all involved loci make an equal contribution to the symptoms of the disease. This should result in a normal curve distribution of genotypes. When it does not, the idea of polygenetic inheritance cannot be supported for that illness.

A Cursory Look at Some Examples

Examples of such diseases are not new to medicine. The above examples are well-known examples of diseases having both genetic and environmental components. Other examples involve atopic diseases such as eczema or dermatitis;spina bifida (open spine), and anencephaly (open skull).

While schizophrenia is widely believed to be multifactorially genetic by biopsychiatrists, no characteristic genetic markers have been determined with any certainty.

If it is shown that the brothers and sisters of the patient have the disease, then there is a strong chance that the disease is genetic and that the patient will also be a genetic carrier. This is not quite enough as it also needs to be proven that the pattern of inheritance is non-Mendelian. This would require studying dozens, even hundreds of different family pedigrees before a conclusion of multifactorial inheritance is drawn. This often takes several years. If multifactorial inheritance is indeed the case, then the chance of the patient contracting the disease is reduced only if cousins and more distant relatives have the disease. It must be stated that while multifactorially-inherited diseases tend to run in families, inheritance will not follow the same pattern as a simple monohybrid or dihybrid cross.

If a genetic cause is suspected and little else is known about the illness, then it remains to be seen exactly how many genes are involved in the phenotypic expression of the disease. Once that is determined, the question must be answered: if two people have the required genes, why are there differences in expression between them? Generally, what makes the two individuals different are likely to be environmental factors. Due to the involved nature of genetic investigations needed to determine such inheritance patterns, this is not usually the first avenue of investigation one would choose to determine etiology.

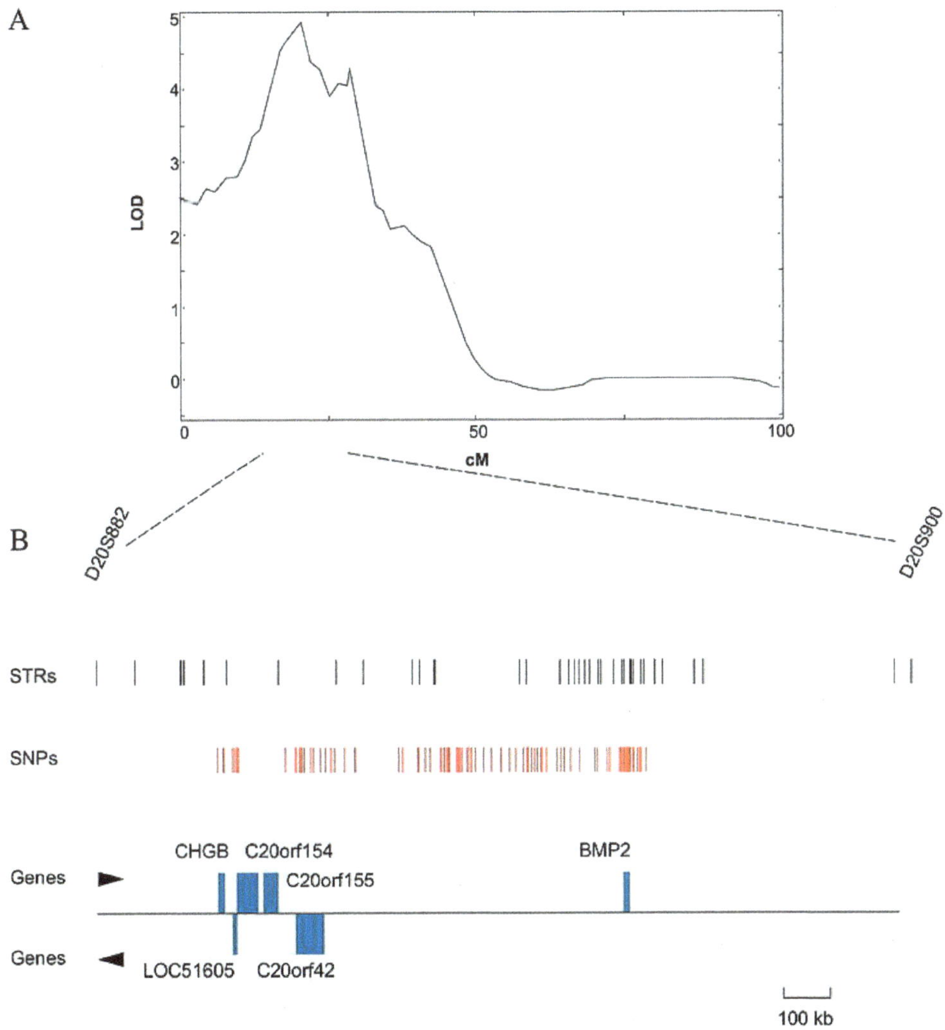

A QTL for osteoporosis on the human chromosome 20.

Typically, QTLs underlie continuous traits (those traits that vary continuously, e.g. height) as opposed to discrete traits (traits that have two or several character values, e.g. red hair in humans, a recessive trait, or smooth vs. wrinkled peas used by Mendel in his experiments).

Moreover, a single phenotypic trait is usually determined by many genes. Consequently, many QTLs are associated with a single trait.

A quantitative trait locus (QTL) is a region of DNA that is associated with a particular phenotypic trait. These QTLs are often found on different chromosomes. Knowing the number of QTLs that explains variation in the phenotypic trait tells us about the genetic architecture of a trait. It may tell us that plant height is controlled by many genes of small effect, or by a few genes of large effect.

Another use of QTLs is to identify candidate genes underlying a trait. Once a region of DNA is identified as contributing to a phenotype, it can be sequenced. The DNA sequence of any genes in this region can then be compared to a database of DNA for genes whose function is already known.

In a recent development, classical QTL analyses are combined with gene expression profiling i.e. by DNA microarrays. Such expression QTLs (eQTLs) describe cis- and trans-controlling elements for the expression of often disease-associated genes. Observed epistatic effects have been found beneficial to identify the gene responsible by a cross-validation of genes within the interacting loci with metabolic pathway- and scientific literature databases.

QTL Mapping

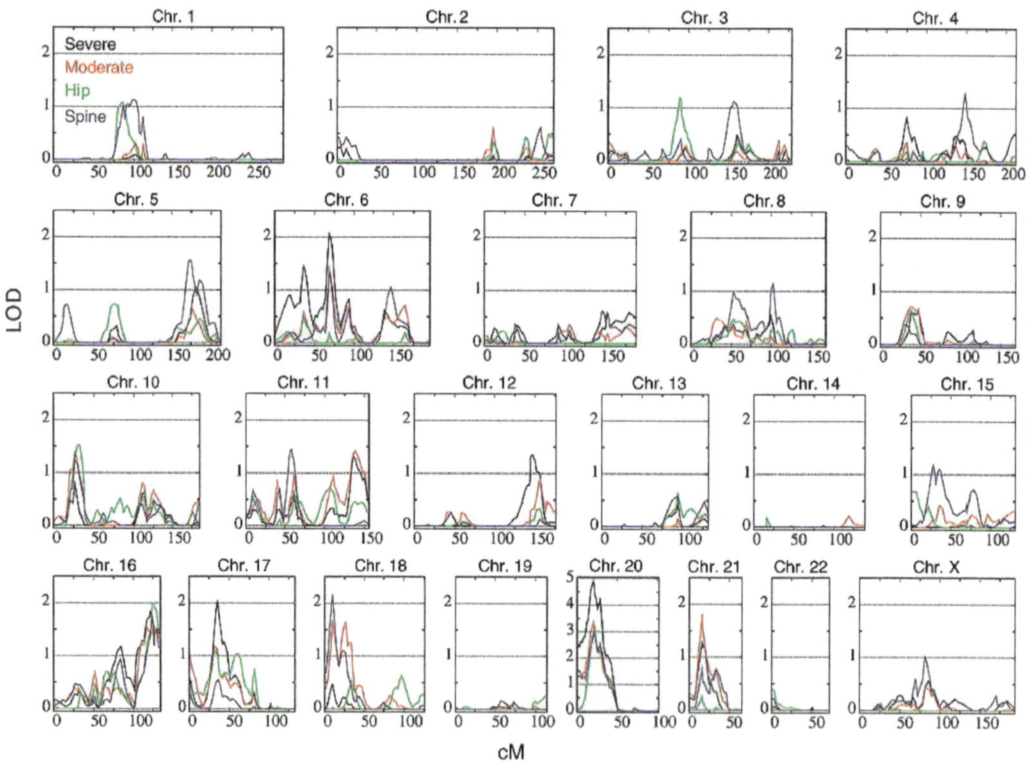

Example of a genome-wide scan for QTL of osteoporosis.

For organisms whose genomes are known, one might now try to exclude genes in the identified region whose function is known with some certainty not to be connected with the trait in question. If the genome is not available, it may be an option to sequence the identified region and determine the putative functions of genes by their similarity to genes with known function, usually in other genomes. This can be done using BLAST, an online tool that allows users to enter a primary sequence and search for similar sequences within the BLAST database of genes from various organisms. It is often not the actual gene underlying the phenotypic trait, but rather a region of DNA that is closely linked with the gene.

Another interest of statistical geneticists using QTL mapping is to determine the complexity of the genetic architecture underlying a phenotypic trait. For example, they may be interested in knowing whether a phenotype is shaped by many independent loci, or by a few loci, and do those loci interact. This can provide information on how the phenotype may be evolving.

Analysis of Variance

The simplest method for QTL mapping is analysis of variance (ANOVA, sometimes called "marker regression") at the marker loci. In this method, in a backcross, one may calculate a t-statistic to compare the averages of the two marker genotype groups. For other types of crosses (such as the intercross), where there are more than two possible genotypes, one uses a more general form of ANOVA, which provides a so-called F-statistic. The ANOVA approach for QTL mapping has three important weaknesses. First, we do not receive separate estimates of QTL location and QTL effect. QTL location is indicated only by looking at which markers give the greatest differences between genotype group averages, and the apparent QTL effect at a marker will be smaller than the true QTL effect as a result of recombination between the marker and the QTL. Second, we must discard individuals whose genotypes are missing at the marker. Third, when the markers are widely spaced, the QTL may be quite far from all markers, and so the power for QTL detection will decrease.

Interval Mapping

Lander and Botstein developed interval mapping, which overcomes the three disadvantages of analysis of variance at marker loci. Interval mapping is currently the most popular approach for QTL mapping in experimental crosses. The method makes use of a genetic map of the typed markers, and, like analysis of variance, assumes the presence of a single QTL. In interval mapping, each locus is considered one at a time and the logarithm of the odds ratio (LOD score) is calculated for the model that the given locus is a true QTL. The odds ratio is related to the Pearson correlation coefficient between the phenotype and the marker genotype for each individual in the experimental cross.

The term 'interval mapping' is used for estimating the position of a QTL within two markers (often indicated as 'marker-bracket'). Interval mapping is originally based on the maximum likelihood but there are also very good approximations possible with simple regression.

The principle for QTL mapping is: 1) The Likelihood can be calculated for a given set of parameters (particularly QTL effect and QTL position) given the observed data on phenotypes and marker genotypes. 2) The estimates for the parameters are those where the likelihood are highest. 3) A significance threshold can be established by permutation testing.

CONVENTIONAL methods for the detection of quantitative trait loci (QTLs) are based on a comparison of single QTL models with a model assuming no QTL. For instance in the "interval mapping" method the likelihood for a single putative QTL is assessed at each location on the genome. However, QTLs located elsewhere on the genome can have an interfering effect. As a consequence, the power of detection may be compromised, and the estimates of locations and effects of QTLs may be biased (LANDER and BOTSTEIN 1989; KNAPP 1991). Even nonexisting so-called "ghost" QTLs may appear (HALEY and KNOTT 1992; MARTINEZ and CURNOW 1992). Therefore, it is obvious that multiple QTLs could be mapped more efficiently and more accurately by using multiple QTL models. One popular approach to handle QTL mapping where multiple QTL contribute to a trait is to iteratively scan the genome and add known QTL to the regression model as QTLs are identified. This method, termed composite interval mapping determine both the location and effects size of QTL more accurately than single-QTL approaches, especially in small mapping populations where the effect of correlation between genotypes in the mapping population may be problematic.

Composite Interval Mapping (CIM)

In this method, one performs interval mapping using a subset of marker loci as covariates. These markers serve as proxies for other QTLs to increase the resolution of interval mapping, by accounting for linked QTLs and reducing the residual variation. The key problem with CIM concerns the choice of suitable marker loci to serve as covariates; once these have been chosen, CIM turns the model selection problem into a single-dimensional scan. The choice of marker covariates has not been solved, however. Not surprisingly, the appropriate markers are those closest to the true QTLs, and so if one could find these, the QTL mapping problem would be complete anyway.

Family-Pedigree Based Mapping

Family based QTL mapping, or Family-pedigree based mapping (Linkage and association mapping), involves multiple families instead of a single family. Family based QTL mapping has been the only way for mapping of genes where experimental crosses are difficult to make. However, due to some advantages, now plant geneticists are attempting to incorporate some of the methods pioneered in human genetics. Using family-pedigree based approach has been discussed (Bink et al. 2008). Family-based linkage and association has been successfully implemented (Rosyara et al. 2009)

MARKER-ASSISTED SELECTION

Marker assisted selection or marker aided selection (MAS) is an indirect selection process where a trait of interest is selected based on a marker (morphological, biochemical or DNA/RNA variation) linked to a trait of interest (e.g. productivity, disease resistance, abiotic stress tolerance, and quality), rather than on the trait itself. This process is used in plant and animal breeding.

For example, using MAS to select individuals with disease resistance involves identifying a marker allele that is linked with disease resistance rather than the level of disease resistance. The assumption is that the marker associates at high frequency with the gene or quantitative trait locus (QTL) of interest, due to genetic linkage (close proximity, on the chromosome, of the marker locus and the disease resistance-determining locus). MAS can be useful to select for traits that are difficult or expensive to measure, exhibit low heritability and/or are expressed late in development. At certain points in the breeding process the specimens are examined to ensure that they express the desired trait.

Marker Types

The majority of MAS work in the present era uses DNA-based markers. However, the first markers that allowed indirect selection of a trait of interest were morphological markers. In 1923, Sax first reported association of a simply inherited genetic marker with a quantitative trait in plants when he observed segregation of seed size associated with segregation for a seed coat color marker in beans (*Phaseolus vulgaris* L.). In 1935, Rasmusson demonstrated linkage of flowering time (a quantitative trait) in peas with a simply inherited gene for flower color.

Markers may be:

- Morphological - These markers are often detectable by eye, by simple visual inspection. Examples of this type of marker include the presence or absence of an awn, leaf sheath coloration, height, grain color, aroma of rice etc. In well-characterized crops like maize, tomato, pea, barley or wheat, tens or hundreds of genes that determine morphological traits have been mapped to specific chromosome locations.
- Biochemical- A protein that can be extracted and observed; for example, isozymes and storage proteins.
- Cytological - The chromosomal banding produced by different stains; for example, G banding.
- DNA-based or molecular- A unique gene (DNA sequence), occurring in proximity to the gene or locus of interest, can be identified by a range of molecular techniques such as RFLP, RAPD, AFLP, DAF, SCAR, microsatellite, or single-nucleotide polymorphism (SNP) detection.

Positive and Negative Selectable Markers

The following terms are generally less relevant to discussions of MAS in plant and animal breeding, but are highly relevant in molecular biology research:

- Positive selectable markers are selectable markers that confer selective advantage to the host organism. An example would be antibiotic resistance, which allows the host organism to survive antibiotic selection.
- Negative selectable markers are selectable markers that eliminate or inhibit growth of the host organism upon selection. An example would be thymidine kinase, which makes the host sensitive to ganciclovir selection.

A distinction can be made between selectable markers (which eliminate certain genotypes from the population) and screenable markers (which cause certain genotypes to be readily identifiable, at which point the experimenter must "score" or evaluate the population and act to retain the preferred genotypes). Most MAS uses screenable markers rather than selectable markers.

Gene vs Marker

The gene of interest directly causes production of protein(s) or RNA that produce a desired trait or phenotype, whereas markers (a DNA sequence or the morphological or biochemical markers produced due to that DNA) are genetically linked to the gene of interest. The gene of interest and the marker tend to move together during segregation of gametes due to their proximity on the same chromosome and concomitant reduction in recombination (chromosome crossover events) between the marker and gene of interest. For some traits, the gene of interest has been discovered and the presence of desirable alleles can be directly assayed with a high level of confidence. However, if the gene of interest is not known, markers linked to the gene of interest can still be used to select for individuals with desirable alleles of the gene of interest. When markers are used there may be some inaccurate results due to inaccurate tests for the marker. There also can be false positive results when markers are used, due to recombination between the marker of interest and gene (or QTL). A perfect marker would elicit no false positive results. The term 'perfect marker' is sometimes used when tests are performed to detect a SNP or other DNA polymorphism in the gene of interest, if that SNP or other polymorphism is the direct cause of the trait of interest. The term 'marker' is still appropriate to use when directly assaying the gene of interest, because the test of genotype is an indirect test of the trait or phenotype of interest.

Important Properties of Ideal Markers for MAS

An ideal marker:

- Easy recognition of all possible phenotypes (homo- and heterozygotes) from all different alleles
- Demonstrates measurable differences in expression between trait types or gene of interest alleles, early in the development of the organism
- Testing for the marker does not have variable success depending on the allele at the marker locus or the allele at the target locus (the gene of interest that determines the trait of interest).
- Low or null interaction among the markers allowing the use of many at the same time in a segregating population
- Abundant in number
- Polymorphic

Demerits of Morphological Markers

Morphological markers are associated with several general deficits that reduce their usefulness including:

- the delay of marker expression until late into the development of the organism
- dominance
- deleterious effects
- pleiotropy
- confounding effects of genes unrelated to the gene or trait of interest but which also affect the morphological marker (epistasis)
- rare polymorphism
- frequent confounding effects of environmental factors which affect the morphological characteristics of the organism

To avoid problems specific to morphological markers, the DNA-based markers have been developed. They are highly polymorphic, exhibit simple inheritance (often codomimant), are abundant throughout the genome, are easy and fast to detect, exhibit minimum pleiotropic effects, and detection is not dependent on the developmental stage of the organism. Numerous markers have been mapped to different chromosomes in several crops including rice, wheat, maize, soybean and several others, and in livestock such as cattle, pigs and chickens. Those markers have been used in diversity analysis, parentage detection, DNA fingerprinting, and prediction of hybrid performance. Molecular markers are useful in indirect selection processes, enabling manual selection of individuals for further propagation.

Selection for Major Genes Linked to Markers

'Major genes' that are responsible for economically important characteristics are frequent in the plant kingdom. Such characteristics include disease resistance, male sterility, self-incompatibility, and others related to shape, color, and architecture of whole plants and are often of mono- or oligogenic in nature. The marker loci that are tightly linked to major genes can be used for selection and are sometimes more efficient than direct selection for the target gene. Such advantages in efficiency may be due for example, to higher expression of the marker mRNA in such cases that the marker is actually a gene. Alternatively, in such cases that the target gene of interest differs between two alleles by a difficult-to-detect single nucleotide polymorphism, an external marker (be it another gene or a polymorphism that is easier to detect, such as a short tandem repeat) may present as the most realistic option.

Situations that are Favorable for Molecular Marker Selection

There are several indications for the use of molecular markers in the selection of a genetic trait.

In such situations that:

- the selected character is expressed late in plant development, like fruit and flower features or adult characters with a juvenile period (so that it is not necessary to wait for the organism to become fully developed before arrangements can be made for propagation)
- the expression of the target gene is recessive (so that individuals which are heterozygous positive for the recessive allele can be crossed to produce some homozygous offspring with the desired trait)
- there is requirement for the presence of special conditions in order to invoke expression of the target gene(s), as in the case of breeding for disease and pest resistance (where inoculation with the disease or subjection to pests would otherwise be required). This advantage derives from the errors due to unreliable inoculation methods and the fact that field inoculation with the pathogen is not allowed in many areas for safety reasons. Moreover, problems in the recognition of the environmentally unstable genes can be eluded.
- the phenotype is affected by two or more unlinked genes (epistatis). For example, selection for multiple genes which provide resistance against diseases or insect pests for gene pyramiding.

The cost of genotyping (an example of a molecular marker assay) is reducing while the cost of phenotyping is increasing particularly in developed countries thus increasing the attractiveness of MAS as the development of the technology continues.

Steps for MAS

Generally the first step is to map the gene or quantitative trait locus (QTL) of interest first by using different techniques and then using this information for marker assisted selection. Generally, the markers to be used should be close to gene of interest (<5 recombination unit or cM) in order to ensure that only minor fraction of the selected individuals will be recombinants. Generally, not only a single marker but rather two markers are used in order to reduce the chances of an error due to homologous recombination. For example, if two flanking markers are used at same time with an interval between them of approximately 20cM, there is higher probability (99%) for recovery of the target gene.

QTL Mapping Techniques

In plants QTL mapping is generally achieved using bi-parental cross populations; a cross between two parents which have a contrasting phenotype for the trait of interest are developed. Commonly used populations are recombinant inbred lines (RILs), doubled haploids (DH), back cross and F_2. Linkage between the phenotype and markers which have already been mapped is tested in these populations in order to determine the position of the QTL. Such techniques are based on linkage and are therefore referred to as "linkage mapping".

Single step MAS and QTL mapping

In contrast to two-step QTL mapping and MAS, a single-step method for breeding typical plant populations has been developed.

In such an approach, in the first few breeding cycles, markers linked to the trait of interest are identified by QTL mapping and later the same information is used in the same population. In this approach, pedigree structure is created from families that are created by crossing number of parents (in three-way or four way crosses). Both phenotyping and genotyping is done using molecular markers mapped the possible location of QTL of interest. This will identify markers and their favorable alleles. Once these favorable marker alleles are identified, the frequency of such alleles will be increased and response to marker assisted selection is estimated. Marker allele(s) with desirable effect will be further used in next selection cycle or other experiments.

High-Throughput Genotyping Techniques

Recently high-throughput genotyping techniques are developed which allows marker aided screening of many genotypes. This will help breeders in shifting traditional breeding to marker aided selection. One example of such automation is using DNA isolation robots, capillary electrophoresis and pipetting robots.

One recent example of capllilary system is Applied Biosystems 3130 Genetic Analyzer. This is the latest generation of 4-capillary electrophoresis instruments for the low to medium throughput laboratories.

Use of MAS for Backcross Breeding

A minimum of five or six-backcross generations are required to transfer a gene of interest from a donor (may not be adapted) to a recipient (recurrent – adapted cultivar). The recovery of the recurrent genotype can be accelerated with the use of molecular markers. If the F1 is heterozygous for the marker locus, individuals with the recurrent parent allele(s) at the marker locus in first or subsequent backcross generations will also carry a chromosome tagged by the marker.

Marker Assisted Gene Pyramiding

Gene pyramiding has been proposed and applied to enhance resistance to disease and insects by selecting for two or more than two genes at a time. For example in rice such pyramids have been developed against bacterial blight and blast. The advantage of use of markers in this case allows to select for QTL-allele-linked markers that have same phenotypic effect.

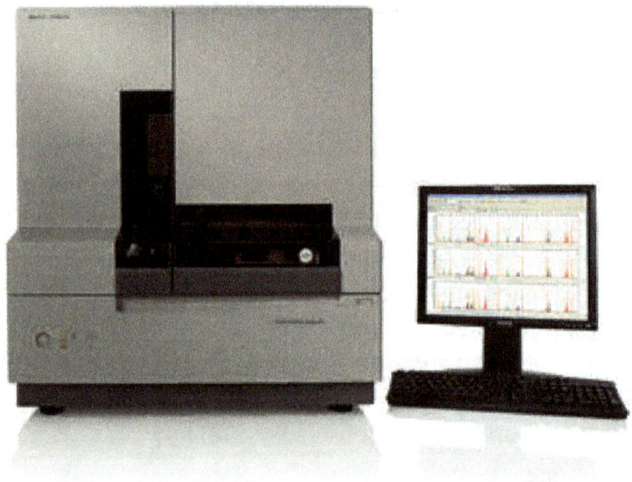

ABI 3130 genetic analyzer.

MAS has also been proved useful for livestock improvement.

A coordinated effort to implement wheat (*Triticum turgidum* and *Triticum aestivum)* marker assisted selection in the U.S. as well as a resource for marker assisted selection exists at the Wheat CAP (Coordinated Agricultural Project) website.

GENETIC ENGINEERING

Genetic engineering, also called genetic modification, is the direct manipulation of an organism's genome using biotechnology. It is a set of technologies used to change the genetic makeup of cells, including the transfer of genes within and across species boundaries to produce improved or novel organisms. New DNA may be inserted in the host genome by first isolating and copying the genetic material of interest using molecular cloning methods to generate a DNA sequence, or by synthesizing the DNA, and then inserting this construct into the host organism. Genes may be removed, or "knocked out", using a nuclease. Gene targeting is a different technique that uses homologous recombination to change an endogenous gene, and can be used to delete a gene, remove exons, add a gene, or introduce point mutations.

An organism that is generated through genetic engineering is considered to be a genetically modified organism (GMO). The first GMOs were bacteria generated in 1973 and GM mice in 1974. Insulin-producing bacteria were commercialized in 1982 and genetically modified food has been sold since 1994. GloFish, the first GMO designed as a pet, was first sold in the United States in December 2003.

Genetic engineering techniques have been applied in numerous fields including research, agriculture, industrial biotechnology, and medicine. Enzymes used in laundry detergent and medicines such as insulin and human growth hormone are now manufactured in GM cells, experimental GM cell lines and GM animals such as mice or zebrafish are being used for research purposes, and genetically modified crops have been commercialized.

Definition

Genetic engineering alters the genetic make-up of an organism using techniques that remove heritable material or that introduce DNA prepared outside the organism either directly into the host or into a cell that is then fused or hybridized with the host. This involves using recombinant nucleic acid (DNA or RNA) techniques to form new combinations of heritable genetic material followed by the incorporation of that material either indirectly through a vector system or directly through micro-injection, macro-injection and micro-encapsulation techniques.

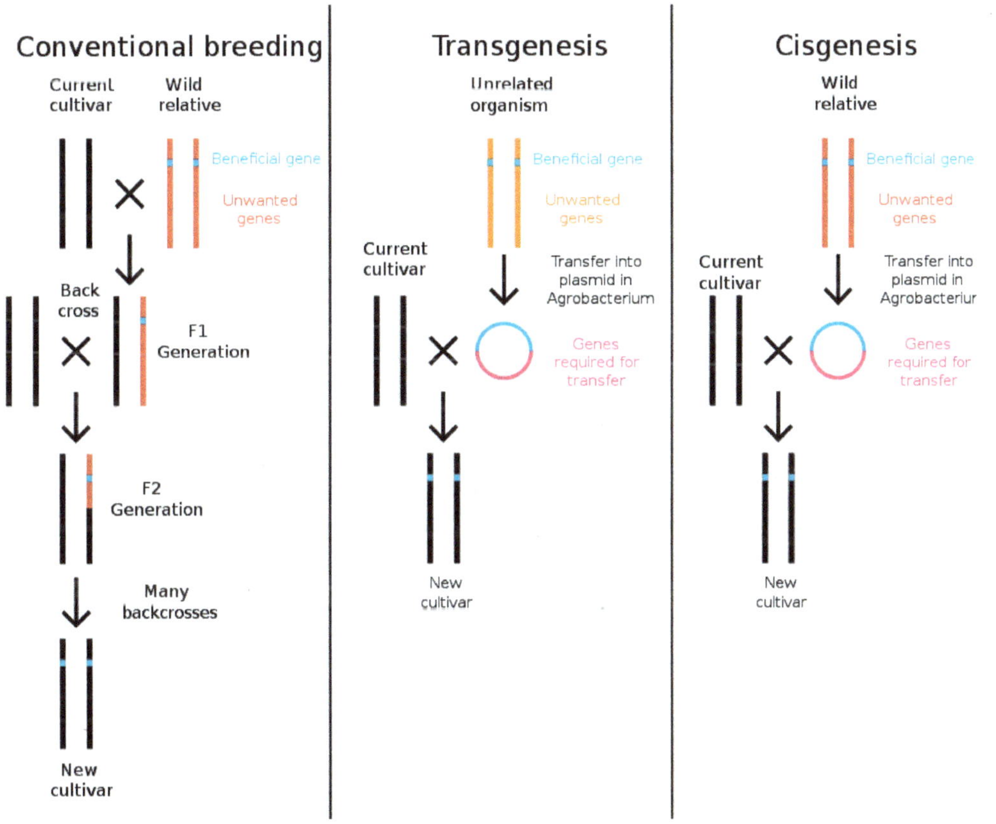

Comparison of conventional plant breeding with transgenic and cisgenic genetic modification.

Genetic engineering does not normally include traditional animal and plant breeding, in vitro fertilization, induction of polyploidy, mutagenesis and cell fusion techniques that do not use recombinant nucleic acids or a genetically modified organism in the process. However the European Commission has also defined genetic engineering broadly as including selective breeding and other means of artificial selection. Cloning and stem cell research, although not considered genetic engineering, are closely related and genetic engineering can be used within them. Synthetic biology is an emerging discipline that takes genetic engineering a step further by introducing artificially synthesized material from raw materials into an organism.

If genetic material from another species is added to the host, the resulting organism is called transgenic. If genetic material from the same species or a species that can naturally breed with the host is used the resulting organism is called cisgenic. Genetic engineering can also be used to remove genetic material from the target organism, creating a gene knockout organism. In Europe genetic modification is synonymous with genetic engineering while within the United States of America it can also refer to conventional breeding methods. The Canadian regulatory system is based on whether a product has novel features regardless of method of origin. In other words, a product is regulated as genetically modified if it carries some trait not previously found in the species whether it was generated using traditional breeding methods (e.g., selective breeding, cell fusion, mutation breeding) or genetic engineering. Within the scientific community, the term *genetic engineering* is not commonly used; more specific terms such as *transgenic* are preferred.

Genetically Modified Organisms

Plants, animals or micro organisms that have changed through genetic engineering are termed genetically modified organisms or GMOs. Bacteria were the first organisms to be genetically modified. Plasmid DNA containing new genes can be inserted into the bacterial cell and the bacteria will then express those genes. These genes can code for medicines or enzymes that process food and other substrates. Plants have been modified for insect protection, herbicide resistance, virus resistance, enhanced nutrition, tolerance to environmental pressures and the production of edible vaccines. Most commercialized GMOs are insect resistant and/or herbicide tolerant crop plants. Genetically modified animals have been used for research, model animals and the production of agricultural or pharmaceutical products.

The genetically modified animals include animals with genes knocked out, increased susceptibility to disease, hormones for extra growth and the ability to express proteins in their milk.

History

Humans have altered the genomes of species for thousands of years through selective breeding, or artificial selection[:1:1] as contrasted with natural selection, and more recently through mutagenesis. Genetic engineering as the direct manipulation of DNA by humans outside breeding and mutations has only existed since the 1970s. The term "genetic engineering" was first coined by Jack Williamson in his science fiction novel *Dragon's Island*, published in 1951 – one year before DNA's role in heredity was confirmed by Alfred Hershey and Martha Chase, and two years before James Watson and Francis Crick showed that the DNA molecule has a double-helix structure – though the general concept of direct genetic manipulation was explored in rudimentary form in Stanley G. Weinbaum's 1936 science fiction story *Proteus Island*.

In 1974 Rudolf Jaenisch created the first GM animal.

In 1972, Paul Berg created the first recombinant DNA molecules by combining DNA from the monkey virus SV40 with that of the lambda virus. In 1973 Herbert Boyer and Stanley Cohen created the first transgenic organism by inserting antibiotic resistance genes into the plasmid of an *E. coli* bacterium. A year later Rudolf Jaenisch created a transgenic mouse by introducing foreign DNA into its embryo, making it the world's first transgenic animal. These achievements led to concerns in the scientific community about potential risks from genetic engineering, which were first discussed in depth at the Asilomar Conference in 1975. One of the main recommendations from this meeting was that government oversight of recombinant DNA research should be established until the technology was deemed safe.

In 1976 Genentech, the first genetic engineering company, was founded by Herbert Boyer and Robert Swanson and a year later the company produced a human protein (somatostatin) in *E.coli*. Genentech announced the production of genetically engineered human insulin in 1978. In 1980, the U.S. Supreme Court in the *Diamond v. Chakrabarty* case ruled that genetically altered life could be patented. The insulin produced by bacteria, branded humulin, was approved for release by the Food and Drug Administration in 1982.

In the 1970s graduate student Steven Lindow of the University of Wisconsin–Madison with D.C. Arny and C. Upper found a bacterium he identified as *P. syringae* that played a role in ice nucleation, and in 1977 he discovered a mutant ice-minus strain. Dr. Lindow (who is now a plant pathologist at the University of California-Berkeley) later successfully created a recombinant ice-minus strain. In 1983, a biotech company, Advanced Genetic Sciences (AGS) applied for U.S. government authorization to perform field tests with the ice-minus strain of *P. syringae* to protect crops from frost, but environmental groups and protestors delayed the field tests for four years with legal challenges. In 1987, the ice-minus strain of *P. syringae* became the first genetically modified organism (GMO) to be released into the environment when a strawberry field and a potato field in California were sprayed with it.Both test fields were attacked by activist groups the night before the tests occurred: "The world's first trial site attracted the world's first field trasher".

The first field trials of genetically engineered plants occurred in France and the USA in 1986, tobacco plants were engineered to be resistant to herbicides. The People's Republic of China was the first country to commercialize transgenic plants, introducing a virus-resistant tobacco in 1992. In 1994 Calgene attained approval to commercially release the Flavr Savr tomato, a tomato engineered to have a longer shelf life. In 1994, the European Union approved tobacco engineered to be resistant to the herbicide bromoxynil, making it the first genetically engineered crop commercialized in Europe. In 1995, Bt Potato was approved safe by the Environmental Protection Agency, after having been approved by the FDA, making it the first pesticide producing crop to be approved in the USA. In 2009 11 transgenic crops were grown commercially in 25 countries, the largest of which by area grown were the USA, Brazil, Argentina, India, Canada, China, Paraguay and South Africa.

In 2010, scientists at the J. Craig Venter Institute created the first synthetic genome and inserted it into an empty bacterial cell. The resulting bacterium, named Synthia, could replicate and produce proteins. In 2014, a bacterium was developed that replicated a plasmid containing a unique base pair, creating the first organism engineered to use an expanded genetic alphabet.

Process

The first step is to choose and isolate the gene that will be inserted into the genetically modified organism. The gene can be isolated using restriction enzymes to cut DNA into fragments and gel electrophoresis to separate them out according to length. Polymerase chain reaction (PCR) can also be used to amplify up a gene segment, which can then be isolated through gel electrophoresis. If the chosen gene or the donor organism's genome has been well studied it may be present in a genetic library. If the DNA sequence is known, but no copies of the gene are available, it can be artificially synthesized.

The gene to be inserted into the genetically modified organism must be combined with other genetic elements in order for it to work properly. The gene can also be modified at this stage for better expression or effectiveness. As well as the gene to be inserted most constructs contain a promoter and terminator region as well as a selectable marker gene. The promoter region initiates transcription of the gene and can be used to control the location and level of gene expression, while the terminator region ends transcription. The selectable marker, which in most cases confers antibiotic resistance to the organism it is expressed in, is needed to determine which cells are transformed with the new gene. The constructs are made using recombinant DNA techniques, such as restriction digests, ligations and molecular cloning. The manipulation of the DNA generally occurs within a plasmid.

The most common form of genetic engineering involves inserting new genetic material randomly within the host genome. Other techniques allow new genetic material to be inserted at a specific location in the host genome or generate mutations at desired genomic loci capable of knocking out endogenous genes. The technique of gene targeting uses homologous recombination to target desired changes to a specific endogenous gene. This tends to occur at a relatively low frequency in plants and animals and generally requires the use of selectable markers. The frequency of gene targeting can be greatly enhanced with the use of engineered nucleases such as zinc finger nucleases, engineered homing endonucleases, or nucleases created from TAL effectors.

In addition to enhancing gene targeting, engineered nucleases can also be used to introduce mutations at endogenous genes that generate a gene knockout.

Transformation

Only about 1% of bacteria are naturally capable of taking up foreign DNA. However, this ability can be induced in other bacteria via stress (e.g. thermal or electric shock), thereby increasing the cell membrane's permeability to DNA; up-taken DNA can either integrate with the genome or exist as extrachromosomal DNA. DNA is generally inserted into animal cells using microinjection, where it can be injected through the cell's nuclear envelope directly into the nucleus or through the use of viral vectors. In plants the DNA is generally inserted using *Agrobacterium*-mediated recombination or biolistics.

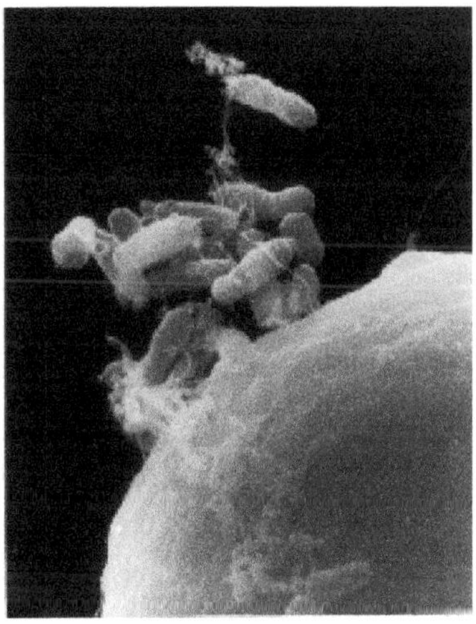

A. tumefaciens attaching itself to a carrot cell.

In *Agrobacterium*-mediated recombination, the plasmid construct contains T-DNA, DNA which is responsible for insertion of the DNA into the host plants genome. This plasmid is transformed into *Agrobacterium* containing no plasmids prior to infecting the plant cells. The *Agrobacterium* will then naturally insert the genetic material into the plant cells. In biolistics transformation particles of gold or tungsten are coated with DNA and then shot into young plant cells or plant embryos. Some genetic material will enter the cells and transform them. This method can be used on plants that are not susceptible to *Agrobacterium* infection and also allows transformation of plant plastids. Another transformation method for plant and animal cells is electroporation. Electroporation involves subjecting the plant or animal cell to an electric shock, which can make the cell membrane permeable to plasmid DNA. In some cases the electroporated cells will incorporate the DNA into their genome. Due to the damage caused to the cells and DNA the transformation efficiency of biolistics and electroporation is lower than agrobacterial mediated transformation and microinjection.

As often only a single cell is transformed with genetic material the organism must be regenerated from that single cell. As bacteria consist of a single cell and reproduce clonally regeneration is not necessary. In plants this is accomplished through the use of tissue culture. Each plant species has different requirements for successful regeneration through tissue culture. If successful an adult plant is produced that contains the transgene in every cell. In animals it is necessary to ensure that the inserted DNA is present in the embryonic stem cells. Selectable markers are used to easily differentiate transformed from untransformed cells. These markers are usually present in the transgenic organism, although a number of strategies have been developed that can remove the selectable marker from the mature transgenic plant. When the offspring is produced they can be screened for the presence of the gene. All offspring from the first generation will be heterozygous for the inserted gene and must be mated together to produce a homozygous animal.

Further testing uses PCR, Southern hybridization, and DNA sequencing is conducted to confirm that an organism contains the new gene. These tests can also confirm the chromosomal location and copy number of the inserted gene. The presence of the gene does not guarantee it will be expressed at appropriate levels in the target tissue so methods that look for and measure the gene products (RNA and protein) are also used. These include northern hybridization, quantitative RT-PCR, Western blot, immunofluorescence, ELISA and phenotypic analysis. For stable transformation the gene should be passed to the offspring in a Mendelian inheritance pattern, so the organism's offspring are also studied.

Genome Editing

Genome editing is a type of genetic engineering in which DNA is inserted, replaced, or removed from a genome using artificially engineered nucleases, or "molecular scissors." The nucleases create specific double-stranded breaks (DSBs) at desired locations in the genome, and harness the cell's endogenous mechanisms to repair the induced break by natural processes of homologous recombination (HR) and nonhomologous end-joining (NHEJ). There are currently four families of engineered nucleases: meganucleases, zinc finger nucleases (ZFNs), transcription activator-like effector nucleases (TALENs), and the Cas9-guideRNA system (adapted from the CRISPR prokarotic immune system). In contrast to artificial genome editing natural genome editing occurs through viral and sub-viral agents competent in identification of genetic syntax structures for insertion/deletion processes with the result of conserved selection processes.

Applications

Genetic engineering has applications in medicine, research, industry and agriculture and can be used on a wide range of plants, animals and micro organisms.

Medicine

In medicine, genetic engineering has been used in manufacturing drugs, to create model animals and do laboratory research, and in gene therapy.

Manufacturing

Genetic engineering is used to mass-produce insulin, human growth hormones, follistim (for treating infertility), human albumin, monoclonal antibodies, antihemophilic factors, vaccines and many other drugs. Mouse hybridomas, cells fused together to create monoclonal antibodies, have been humanised through genetic engineering to create human monoclonal antibodies. Genetically engineered viruses are being developed that can still confer immunity, but lack the infectious sequences.

Research

Genetic engineering is used to create animal models of human diseases. Genetically modified mice are the most common genetically engineered animal model. They have been used to study and model cancer (the oncomouse), obesity, heart disease, diabetes, arthritis, substance abuse, anxiety, aging and Parkinson disease. Potential cures can be tested against these mouse models. Also genetically modified pigs have been bred with the aim of increasing the success of pig to human organ transplantation.

Gene Therapy

Gene therapy is the genetic engineering of humans, generally by replacing defective genes with effective ones. This can occur in somatic tissue or germline tissue.

Somatic gene therapy has been studied in clinical research in several diseases, including X-linked SCID, chronic lymphocytic leukemia (CLL), and Parkinson's disease. In 2012, Glybera became the first gene therapy treatment to be approved for clinical use in either Europe or the United States after its endorsement by the European Commission.

With regard to germline gene therapy, the scientific community has been opposed to attempts to alter genes in humans in inheritable ways using biotechnology since the technology was first introduced, and the caution has continued as the technology has progressed. With the advent of new techniques like CRISPR, in March 2015 scientists urged a worldwide ban on clinical use of gene editing technologies to edit the human genome in a way that can be inherited. In April 2015, Chinese researchers sparked controversy when they reported results of basic research experiments in which they edited the DNA of non-viable human embryos using CRISPR. In December 2015, scientists of major world academies called for a moratorium on inheritable human genome edits, including those related to CRISPR-Cas9 technologies.

There are also ethical concerns should the technology be used not just for treatment, but for enhancement, modification or alteration of a human beings' appearance, adaptability, intelligence, character or behavior. The distinction between cure and enhancement can also be difficult to establish. Transhumanists consider the enhancement of humans desirable.

Research

Genetic engineering is an important tool for natural scientists. Genes and other genetic information from a wide range of organisms are transformed into bacteria for storage and modification, creating genetically modified bacteria in the process. Bacteria are cheap, easy to grow, clonal, multiply quickly, relatively easy to transform and can be stored at -80 °C almost indefinitely. Once a gene is isolated it can be stored inside the bacteria providing an unlimited supply for research.

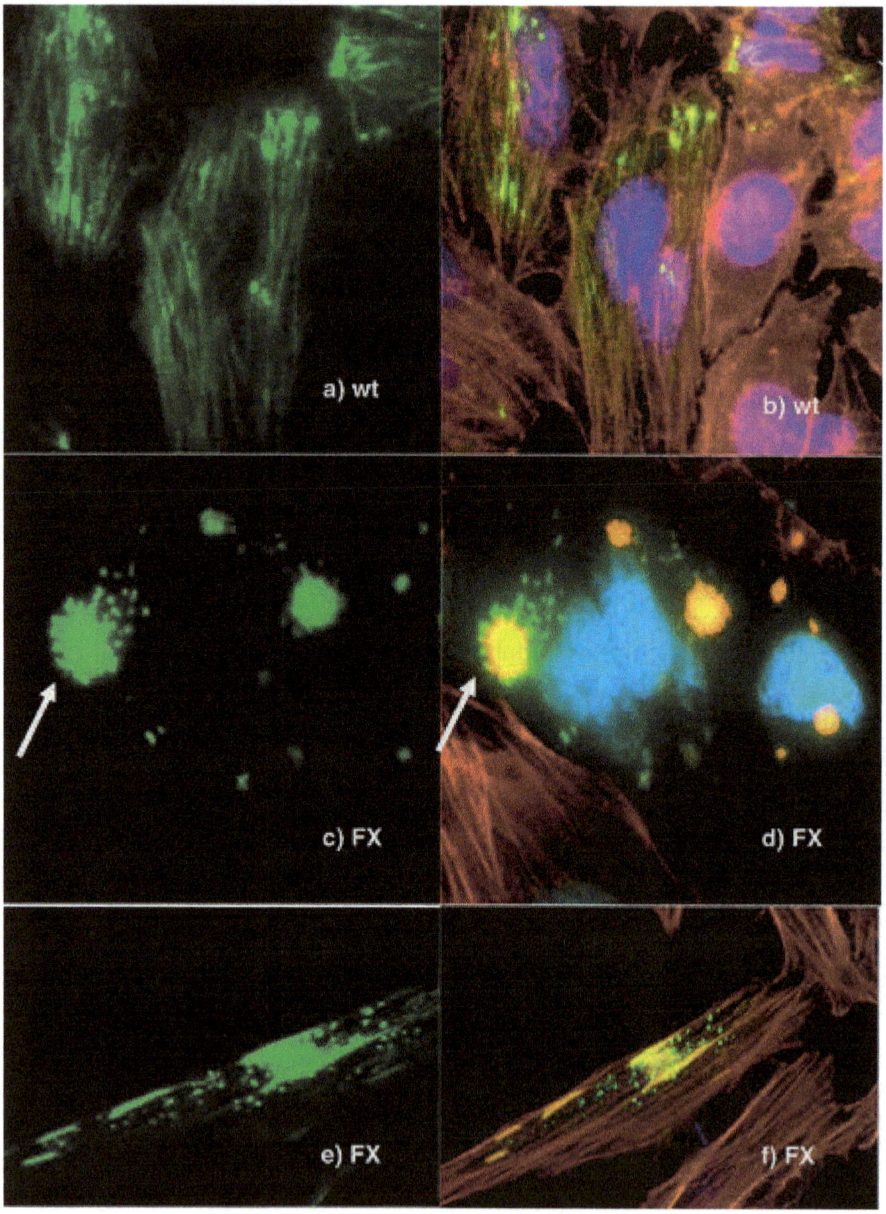

Human cells in which some proteins are fused with green fluorescent protein to allow them to be visualized.

Organisms are genetically engineered to discover the functions of certain genes. This could be the effect on the phenotype of the organism, where the gene is expressed or what other genes it interacts with. These experiments generally involve loss of function, gain of function, tracking and expression.

- Loss of function experiments, such as in a gene knockout experiment, in which an organism is engineered to lack the activity of one or more genes. A knockout experiment involves the creation and manipulation of a DNA construct *in vitro*, which, in a simple knockout, consists of a copy of the desired gene, which has been altered such that it is non-functional. Embryonic stem cells incorporate the altered gene, which replaces the already present functional copy. These stem cells are injected into blastocysts, which are implanted into surrogate mothers. This allows the experimenter to analyze the defects caused by this mutation and thereby determine the role of particular genes. It is used especially frequently in developmental biology. Another method, useful in organisms such as Drosophila (fruit fly), is to induce mutations in a large population and then screen the progeny for the desired mutation. A similar process can be used in both plants and prokaryotes. Loss of function tells whether or not a protein is required for a function, but it does not always mean it's sufficient, especially if a function requires multiple proteins and lose the said function if one protein is missing.
- Gain of function experiments, the logical counterpart of knockouts. These are sometimes performed in conjunction with knockout experiments to more finely establish the function of the desired gene. The process is much the same as that in knockout engineering, except that the construct is designed to increase the function of the gene, usually by providing extra copies of the gene or inducing synthesis of the protein more frequently. Gain of function is used to tell whether or not a protein is sufficient for a function, but it does not always mean it's required. Especially when dealing with genetic/functional redundancy.
- Tracking experiments, which seek to gain information about the localization and interaction of the desired protein. One way to do this is to replace the wild-type gene with a 'fusion' gene, which is a juxtaposition of the wild-type gene with a reporting element such as green fluorescent protein (GFP) that will allow easy visualization of the products of the genetic modification. While this is a useful technique, the manipulation can destroy the function of the gene, creating secondary effects and possibly calling into question the results of the experiment. More sophisticated techniques are now in development that can track protein products without mitigating their function, such as the addition of small sequences that will serve as binding motifs to monoclonal antibodies.
- Expression studies aim to discover where and when specific proteins are produced. In these experiments, the DNA sequence before the DNA that codes for a protein, known as a gene's promoter, is reintroduced into an organism with the protein coding region replaced by a reporter gene such as GFP or an enzyme that catalyzes the production of a dye. Thus the time and place where a particular protein is produced can be observed. Expression studies can be taken a step further by altering the promoter to find which pieces are crucial for the proper expression of the gene and are actually bound by transcription factor proteins; this process is known as promoter bashing.

Industrial

Using genetic engineering techniques one can transform microorganisms such as bacteria or yeast, or transform cells from multicellular organisms such as insects or mammals, with a gene coding for a useful protein, such as an enzyme, so that the transformed organism will overexpress the desired protein. One can manufacture mass quantities of the protein by growing the transformed organism in bioreactor equipment using techniques of industrial fermentation, and then purifying the protein. Some genes do not work well in bacteria, so yeast, insect cells, or mammalians cells, each a eukaryote, can also be used. These techniques are used to produce medicines such as insulin, human growth hormone, and vaccines, supplements such as tryptophan, aid in the production of food (chymosin in cheese making) and fuels. Other applications involving genetically engineered bacteria being investigated involve making the bacteria perform tasks outside their natural cycle, such as making biofuels, cleaning up oil spills, carbon and other toxic waste and detecting arsenic in drinking water.Certain genetically modified microbes can also be used in biomining and bioremediation, due to their ability to extract heavy metals from their environment and incorporate them into compounds that are more easily recoverable.

Experimental, Lab Scale Industrial Applications

In materials science, a genetically modified virus has been used in an academic lab as a scaffold for assembling a more environmentally friendly lithium-ion battery.

Bacteria have been engineered to function as sensors by expressing a fluorescent protein under certain environmental conditions.

Agriculture

One of the best-known and controversial applications of genetic engineering is the creation and use of genetically modified crops or genetically modified organisms, such as genetically modified fish, which are used to produce genetically modified food and materials with diverse uses. There are four main goals in generating genetically modified crops.

One goal, and the first to be realized commercially, is to provide protection from environmental threats, such as cold (in the case of Ice-minus bacteria), or pathogens, such as insects or viruses, and/or resistance to herbicides. There are also fungal and virus resistant crops developed or in development. They have been developed to make the insect and weed management of crops easier and can indirectly increase crop yield.

Another goal in generating GMOs is to modify the quality of produce by, for instance, increasing the nutritional value or providing more industrially useful qualities or quantities. The Amflora potato, for example, produces a more industrially useful blend of starches. Cows have been engineered to produce more protein in their milk to facilitate cheese production. Soybeans and canola have been genetically modified to produce more healthy oils.

Bt-toxins present in peanut leaves (bottom image) protect it from extensive damage caused by European corn borer larvae (top image).

Another goal consists of driving the GMO to produce materials that it does not normally make. One example is "pharming", which uses crops as bioreactors to produce vaccines, drug intermediates, or drug themselves; the useful product is purified from the harvest and then used in the standard pharmaceutical production process. Cows and goats have been engineered to express drugs and other proteins in their milk, and in 2009 the FDA approved a drug produced in goat milk.

Another goal in generating GMOs, is to directly improve yield by accelerating growth, or making the organism more hardy (for plants, by improving salt, cold or drought tolerance).Salmon have been genetically modified with growth hormones to increase their size.

The genetic engineering of agricultural crops can increase the growth rates and resistance to different diseases caused by pathogens and parasites. This is beneficial as it can greatly increase the production of food sources with the usage of fewer resources that would be required to host the world's growing populations. These modified crops would also reduce the usage of chemicals, such as fertilizers and pesticides, and therefore decrease the severity and frequency of the damages produced by these chemical pollution.

There is a scientific consensus that currently available food derived from GM crops poses no greater risk to human health than conventional food,but that each GM food needs to be tested on a case-by-case basis before introduction. Nonetheless, members of the public are much less likely than scientists to perceive GM foods as safe. The legal and regulatory status of GM foods varies by country, with some nations banning or restricting them, and others permitting them with widely differing degrees of regulation. Gene flow into related non-transgenic crops, off target effects on beneficial organisms and the impact on biodiversity are important environmental issues. Ethical concerns involve religious issues, corporate control of the food supply, intellectual property rights and the level of labeling needed on genetically modified products.

Conservation

Genetic engineering has potential applications in conservation and natural areas management. For example, gene transfer through viral vectors has been proposed as a means of controlling invasive species as well as vaccinating threatened fauna from disease. Transgenic trees have also been suggested as a way to confer resistance to pathogens in wild populations. With the increasing risks of maladaptation in organisms as a result of climate change and other perturbations, facilitated adaptation through gene tweaking could be one solution to reducing extinction risks. Applications of genetic engineering in conservation are thus far mostly theoretical and have yet to be put into practice. Further experimentation will be necessary to gauge the benefits and costs of such practices.

BioArt and Entertainment

Genetic engineering is also being used to create BioArt. Some bacteria have been genetically engineered to create black and white photographs.

Genetic engineering has also been used to create novelty items such as lavender-colored carnations, blue roses, and glowing fish.

Regulation

The regulation of genetic engineering concerns the approaches taken by governments to assess and manage the risks associated with the development and release of genetically modified crops. There are differences in the regulation of GM crops between countries, with some of the most marked differences occurring between the USA and Europe. Regulation varies in a given country depending on the intended use of the products of the genetic engineering. For example, a crop not intended for food use is generally not reviewed by authorities responsible for food safety. Starting in the late 1980s, guidance on assessing the safety of genetically engineered plants and food emerged from organizations including the FAO and WHO.

Controversy

Critics have objected to use of genetic engineering per se on several grounds, including ethical concerns, ecological concerns, and economic concerns raised by the fact GM techniques and GM organisms are subject to intellectual property law. GMOs also are involved in controversies over GM food with respect to whether food produced from GM crops is safe, whether it should be labeled, and whether GM crops are needed to address the world's food needs. See the genetically modified food controversies article for discussion of issues about GM crops and GM food. These controversies have led to litigation, international trade disputes, and protests, and to restrictive regulation of commercial products in some countries.

REFERENCES

1. Voosen P (2009) Molecular Breeding Makes Crops Hardier and More Nutritious Markers, knockouts and other technical advances improve breeding without modifying genes, Scientific American
2. Stephen P. Moose* and Rita H. Mumm (2008) Molecular Plant Breeding as the Foundation for 21st Century Crop Improvement, Plant Physiology 147:969-977
3. Jannink, Jean-Luc; Lorenz, Aaron J.; Iwata, Hiroyoshi (2010-03-01). "Genomic selection in plant breeding: from theory to practice". Briefings in Functional Genomics. 9 (2): 166–177. doi:10.1093/bfgp/elq001. ISSN 2041-2649. PMID 20156985.
4. Heffner, Elliot L.; Sorrells, Mark E.; Jannink, Jean-Luc (2009-01-01). "Genomic Selection for Crop Improvement". Crop Science. 49 (1). doi:10.2135/cropsci2008.08.0512. ISSN 1435-0653.
5. Goddard, ME; Hayes, BJ (2007). "Genomic selection". Journal of animal breeding and genetics = Zeitschrift fur Tierzuchtung und Zuchtungsbiologie. 124 (6): 323–30. doi:10.1111/j.1439-0388.2007.00702.x. PMID 18076469.
6. Miles, C; Wayne, M (2008). "Quantitative trait locus (QTL) analysis". Nature Education (1.1).
7. http://www.literature.org/authors/darwin-charles/the-origin-of-species/
8. Castle WE (1903). "MENDEL'S LAW OF HEREDITY". Science. 18: 396–406. doi:10.1126/science.18.456.396. PMID 17752783.
9. http://www.genetics.org/content/36/3/254.long
10. http://www.genetics.org/content/16/2/97.long
11. Ricki Lewis (2003), Multifactorial Traits, McGraw-Hill Higher Education.
12. Proud, Virginia & Roberts, Helen (31 December 2005). "Medical Genetics: Multifactorial Inheritance". Children's Hospital of the King's Daughters. Retrieved 6 January 2007.
13. "Multifactorial Inheritance". Pregnancy and Newborn Health Education Centre. The March of Dimes. Archived from the original on 2 November 2006. Retrieved November 12, 2014.
14. Emery's Elements of Medical Genetics
15. Tissot, Robert. "Human Genetics for 1st Year Students: Multifactorial Inheritance". Retrieved 6 January 2007.

16. Birth Defects Genetics Centre, University of South Dakota School of Medicine. "Multifactorial Inheritance". Clinical Genetics: A Self-Study Guide for Health Care Providers. University of South Dakota School of Medicine. Archived from the original on 30 December 2006. Retrieved 6 January 2007.
17. "Definition of Multifactorial inheritance". MedicineNet.com MedTerms Dictionary. MedicineNet.com. Retrieved 6 January 2007.
18. Turnpenny, Peter (2004). "Emery's Elements of Medical Genetics" (PDF) (12th ed.). Elsevier. Retrieved 6 January 2007. |chapter= ignored (help)
19. Westra HJ, et al. (2013). "Systematic identification of trans eQTLs as putative drivers of known disease associations". Nat Genet. 45 (10): 1238–1243. doi:10.1038/ng.2756.
20. Lynch, M. & Walsh, B. Genetics and Analysis of Quantitative Traits edn 1 (Sinauer Associates, 1998).
21. Bloom J. S.; Ehrenreich I. M.; Loo W. T.; Lite T.-L. V.; Kruglyak L. (2013). "Finding the sources of missing heritability in a yeast cross". Nature. 494 (7436): 234–237. doi:10.1038/nature11867.
22. Mapping Mendelian factors underlying quantitative traits using RFLP linkage maps. ES Lander and D Botstein. Genetics. 1989
23. http://www.genetics.org/content/135/1/205.full.pdf
24. Jannink, J; Bink, Mc; Jansen, Rc (Aug 2001). "Using complex plant pedigrees to map valuable genes". Trends in Plant Science. 6 (8): 337–42. doi:10.1016/S1360-1385(01)02017-9. ISSN 1360-1385. PMID 11495765.
25. Rosyara, U. R.; Maxson-stein, K.L.; Glover, K.D.; Stein, J.M.; Gonzalez-hernandez, J.L. (2007). "Family-based mapping of FHB resistance QTLs in hexaploid wheat". Proceedings of National Fusarium head blight forum.
26. Bink MCAM, Boer MP, ter Braak CJF, Jansen J, Voorrips RE, van de Weg WE: Bayesian analysis of complex traits in pedigreed plant populations.
27. Euphytica 2008, 161:85–96.
28. Rosyara U.R., J.L. Gonzalez-Hernandez, K.D. Glover, K.R. Gedye and J.M. Stein. 2009. Family-based mapping of quantitative trait loci in plant breeding populations with resistance to Fusarium head blight in wheat as an illustration Theoretical Applied Genetics 118:1617–1631
29. Garnier, Sophie, Truong, Vinh, Genome-Wide Haplotype Analysis of Cis Expression Quantitative Trait Loci in Monocytes
30. Ribaut, J.-M. et al., Genetic basis of physiological traits. In Application of Physiology in Wheat Breeding, CIMMYT, Mexico, 2001.
31. Ribaut, J.-M. and Hoisington, D. A., Marker assisted selection: new tools and strategies. Trends Plant Sci., 1998, 3, 236–239.
32. Rosyara, U.R. 2006. REQUIREMENT OF ROBUST MOLECULAR MARKER TECHNOLOGY FOR PLANT BREEDING APPLICATIONS.Journal of Plant Breed. Gr. 1: 67 – 72. click to download

33. "positive selection". Scitable. Nature. Retrieved 29 September 2011.

34. "negative selection". Scitable. Nature. Retrieved 29 September 2011.

35. Nowicki, Marcin; et al. (26 October 2013), More than meets the eye: A multi-year expressivity analyses of tomato sterility in ps and ps-2 lines (PDF), Australian Journal of Crop Science, 7(13):2154-2161; Southern Cross Publishing, retrieved 2013-10-29

36. Rosyara, U. R.; K.L. Maxson-Stein; K.D. Glover; J.M. Stein; J.L. Gonzalez-Hernandez. 2007. Family-based mapping of FHB resistance QTLs in hexaploid wheat. Proceedings of National Fusarium head blight forum, 2007, Dec 2-4, Kansas City, MO.

37. Rosyara U.R., J.L. Gonzalez-Hernandez, K.D. Glover, K.R. Gedye and J.M. Stein. 2009. Family-based mapping of quantitative trait loci in plant breeding populations with resistance to Fusarium head blight in wheat as an illustration Theoretical Applied Genetics 118:1617-1631

38. Dekkers., J. C. M. 2004. Commercial application of marker- and gene-assisted selection in livestock: Strategies and lessons. J. Anim. Sci. 82:E313-E328.

39. "First transgenic pet, 'GloFish', sold to US public". PHG Foundation. 9 January 2004.

40. "Terms and Acronyms". U.S. Environmental Protection Agency online. Retrieved 16 July 2015.

41. Vert, Michel; Doi, Yoshiharu; Hellwich, Karl-Heinz; Hess, Michael; Hodge, Philip; Kubisa, Przemyslaw; Rinaudo, Marguerite; Schué, François (2012). "Terminology for biorelated polymers and applications (IUPAC Recommendations 2012)" (PDF). Pure and Applied Chemistry. 84 (2): 377–410. doi:10.1351/PAC-REC-10-12-04.

42. The European Parliament and the council of the European Union (12 March 2001). "Directive on the release of genetically modified organisms (GMOs) Directive 2001/18/EC ANNEX I A". Official Journal of the European Communities: 17.

43. Staff Economic Impacts of Genetically Modified Crops on the Agri-Food Sector; P. 42 Glossary - Term and Definitions The European Commission Directorate-General for Agriculture, "Genetic engineering: The manipulation of an organism's genetic endowment by introducing or eliminating specific genes through modern molecular biology techniques. A broad definition of genetic engineering also includes selective breeding and other means of artificial selection.", Retrieved 5 November 2012

44. Van Eenennaam, Alison. "Is Livestock Cloning Another Form of Genetic Engineering?" (PDF). agbiotech. Archived from the original (PDF) on 11 May 2011.

45. Suter, David M.; Dubois-Dauphin, Michel; Krause, Karl-Heinz (2006). "Genetic engineering of embryonic stem cells" (PDF). Swiss Med Wkly. 136 (27–28): 413–415. PMID 16897894.

46. Andrianantoandro, Ernesto; Basu, Subhayu; Kariga, David K.; Weiss, Ron (16 May 2006). "Synthetic biology: new engineering rules for an emerging discipline". Molecular Systems Biology. 2 (2006.0028): 2006.0028. doi:10.1038/msb4100073. PMC 1681505Freely accessible. PMID 16738572.
47. Jacobsen, E.; Schouten, H. J. (2008). "Cisgenesis, a New Tool for Traditional Plant Breeding, Should be Exempted from the Regulation on Genetically Modified Organisms in a Step by Step Approach". Potato Research. 51: 75–88. doi:10.1007/s11540-008-9097-y.
48. Capecchi, Mario R. (2001). "Generating mice with targeted mutations". Nature Medicine. 7 (10): 1086–90. doi:10.1038/nm1001-1086. PMID 11590420.
49. Staff Biotechnology - Glossary of Agricultural Biotechnology Terms United States Department of Agriculture, "Genetic modification: The production of heritable improvements in plants or animals for specific uses, via either genetic engineering or other more traditional methods. Some countries other than the United States use this term to refer specifically to genetic engineering.", Retrieved 5 November 2012
50. Maryanski, James H. (19 October 1999). "Genetically Engineered Foods". Center for Food Safety and Applied Nutrition at the Food and Drug Administration.
51. Evans, Brent and Lupescu, Mihai (15 July 2012) Canada - Agricultural Biotechnology Annual – 2012 GAIN (Global Agricultural Information Network) report CA12029, United States Department of Agriculture, Foreifn Agricultural Service, Retrieved 5 November 2012
52. McHugen, Alan (14 September 2000). "Chapter 1: Hors-d'oeuvres and entrees/What is genetic modification? What are GMOs?". Pandora's Picnic Basket. Oxford University Press. ISBN 978-0198506744.
53. Staff (28 November 2005) Health Canada - The Regulation of Genetically Modified Food Glossary definition of Genetically Modified: "An organism, such as a plant, animal or bacterium, is considered genetically modified if its genetic material has been altered through any method, including conventional breeding. A 'GMO' is a genetically modified organism.", Retrieved 5 November 2012
54. "What is genetic modification (GM)?". CSIRO.
55. "Genetic Modification of Bacteria". Annenberg Foundation.
56. Panesar, Pamit et al (2010) "Enzymes in Food Processing: Fundamentals and Potential Applications", Chapter 10, I K International Publishing House, ISBN 978-9380026336
57. "GM traits list". International Service for the Acquisition of Agri-Biotech Applications.
58. "ISAAA Brief 43-2011: Executive Summary". International Service for the Acquisition of Agri-Biotech Applications.
59. Connor, Steve (2 November 2007). "The mouse that shook the world". The Independent.

60. Root, Clive (2007). Domestication. Greenwood Publishing Groups.

61. Zohary, Daniel; Hopf, Maria; Weiss, Ehud (2012). Domestication of Plants in the Old World: The origin and spread of plants in the old world. Oxford University Press.

62. Stableford, Brian M. (2004). Historical dictionary of science fiction literature. p. 133. ISBN 9780810849389.

63. A, Hershey; Chase, M. (1952). "Independent functions of viral protein and nucleic acid in growth of bacteriophage" (PDF). J Gen Physiol. 36 (1): 39–56. doi:10.1085/jgp.36.1.39. PMC 2147348Freely accessible. PMID 12981234.

64. "Genetic Engineering". Encyclopedia of Science Fiction. April 2, 2015.

65. Shiv Kant Prasad, Ajay Dash (2008). Modern Concepts in Nanotechnology, Volume 5. Discovery Publishing House. ISBN 9788183562966.

66. Jackson, DA; Symons, RH; Berg, P (1 October 1972). "Biochemical Method for Inserting New Genetic Information into DNA of Simian Virus 40: Circular SV40 DNA Molecules Containing Lambda Phage Genes and the Galactose Operon of Escherichia coli". PNAS. 69 (10): 2904–2909. Bibcode:1972PNAS...69.2904J. doi:10.1073/pnas.69.10.2904. PMC 389671Freely accessible. PMID 4342968.

67. Arnold, Paul (2009). "History of Genetics: Genetic Engineering Timeline".

68. Cohen, Stanley N.; Chang, Annie C. Y. (1 May 1973). "Recircularization and Autonomous Replication of a Sheared R-Factor DNA Segment in Escherichia coli Transformants — PNAS". Pnas.org. Retrieved 17 July 2010.

69. Jaenisch, R. and Mintz, B. (1974) Simian virus 40 DNA sequences in DNA of healthy adult mice derived from preimplantation blastocysts injected with viral DNA. Proc. Natl. Acad. 71(4) 1250–1254

70. Berg P; Baltimore, D; Brenner, S; Roblin, RO; Singer, MF; et al. (1975). "Summary statement of the Asilomar Conference on recombinant DNA molecules" (PDF). Proc. Natl. Acad. Sci. U.S.A. 72 (6): 1981–4. Bibcode:1975PNAS...72.1981B. doi:10.1073/pnas.72.6.1981. PMC 432675Freely accessible. PMID 806076.

71. NIH Guidelines for research involving recombinant DNA molecules Archived 10 September 2012 at the Wayback Machine.

72. Goeddel, David; Kleid, Dennis G.; Bolivar, Francisco; Heyneker, Herbert L.; Yansura, Daniel G.; Crea, Roberto; Hirose, Tadaaki; Kraszewski, Adam; Itakura, Keiichi; Riggs, Arthur D. (January 1979). "Expression in Escherichia coli of chemically synthesized genes for human insulin" (PDF). PNAS. 76 (1): 106–110. Bibcode:1979PNAS...76..106G. doi:10.1073/pnas.76.1.106. PMC 382885Freely accessible. PMID 85300.

73. US Supreme Court Cases from Justia & Oyez (16 June 1980). "Diamond V Chakrabarty". 447 (303). Supreme.justia.com. Retrieved 17 July 2010.

74. "Artificial Genes". TIME. 15 November 1982. Retrieved 17 July 2010.

75. H. Patricia Hynes. (1989) Biotechnology in agriculture: an analysis of selected technologies and policy in the United States. Reproductive and Genetic Engineering (2)1:39–49

76. Bratspies, Rebecca (2007). "Some Thoughts on the American Approach to Regulating Genetically Modified Organisms" (PDF). Kansas Journal of Law & Public Policy. 16 (3): 101–131.

77. BBC News 14 June 2002 GM crops: A bitter harvest?

78. Thomas H. Maugh II for the Los Angeles Times. 9 June 1987. Altered Bacterium Does Its Job : Frost Failed to Damage Sprayed Test Crop, Company Says

79. James, Clive (1996). "Global Review of the Field Testing and Commercialization of Transgenic Plants: 1986 to 1995" (PDF). The International Service for the Acquisition of Agri-biotech Applications. Retrieved 17 July 2010.

80. James, Clive (1997). "Global Status of Transgenic Crops in 1997" (PDF). ISAAA Briefs No. 5.: 31.

81. Bruening, G.; Lyons, J.M. (2000). "The case of the FLAVR SAVR tomato". California Agriculture. 54 (4): 6–7. doi:10.3733/ca.v054n04p6.

82. MacKenzie, Debora (18 June 1994). "Transgenic tobacco is European first". New Scientist.

83. Genetically Altered Potato Ok'd For Crops Lawrence Journal-World - 6 May 1995

84. Global Status of Commercialized Biotech/GM Crops: 2009 ISAAA Brief 41-2009, 23 February 2010. Retrieved 10 August 2010

85. Pennisi, Elizabeth (2010-05-21). "Synthetic Genome Brings New Life to Bacterium". Science. 328 (5981): 958–959. doi:10.1126/science.328.5981.958. ISSN 0036-8075. PMID 20488994.

86. Gibson, D. G.; Glass, J. I.; Lartigue, C.; Noskov, V. N.; Chuang, R.-Y.; Algire, M. A.; Benders, G. A.; Montague, M. G.; Ma, L.; Moodie, M. M.; Merryman, C.; Vashee, S.; Krishnakumar, R.; Assad-Garcia, N.; Andrews-Pfannkoch, C.; Denisova, E. A.; Young, L.; Qi, Z.-Q.; Segall-Shapiro, T. H.; Calvey, C. H.; Parmar, P. P.; Hutchison Ca, C. A.; Smith, H. O.; Venter, J. C. (2010). "Creation of a Bacterial Cell Controlled by a Chemically Synthesized Genome". Science. 329 (5987): 52–6. doi:10.1126/science.1190719. PMID 20488990.

87. Malyshev, Denis A.; Dhami, Kirandeep; Lavergne, Thomas; Chen, Tingjian; Dai, Nan; Foster, Jeremy M.; Corrêa, Ivan R.; Romesberg, Floyd E. (2014-05-15). "A semi-synthetic organism with an expanded genetic alphabet". Nature. 509 (7500): 385–388. doi:10.1038/nature13314. ISSN 0028-0836. PMC 4058825Freely accessible. PMID 24805238.

88. Thyer, Ross; Ellefson, Jared (2014-05-15). "Synthetic biology: New letters for life's alphabet". Nature. 509 (7500): 291–292. doi:10.1038/nature13335. ISSN 0028-0836.

89. Alberts B, Johnson A, Lewis J, et al. (2002). "8". Isolating, Cloning, and Sequencing DNA. (4th ed.). New York: Garland Science.

90. Kaufman, R I; Nixon, B T (1996). "Use of PCR to isolate genes encoding sigma54-dependent activators from diverse bacteria". J Bacteriol. 178 (13): 3967–3970. doi:10.1128/jb.178.13.3967-3970.1996. PMC 232662Freely accessible. PMID 8682806.

91. Liang, Jing; Luo, Yunzi; Zhao, Huimin (2011). "Synthetic biology: Putting synthesis into biology". Wiley Interdisciplinary Reviews: Systems Biology and Medicine. 3: 7–20. doi:10.1002/wsbm.104.

92. Berg, P.; Mertz, J. E. (2010). "Personal Reflections on the Origins and Emergence of Recombinant DNA Technology". Genetics. 184 (1): 9–17. doi:10.1534/genetics.109.112144. PMC 2815933Freely accessible. PMID 20061565.

93. Townsend JA, Wright DA, Winfrey RJ, et al. (May 2009). "High-frequency modification of plant genes using engineered zinc-finger nucleases". Nature. 459 (7245): 442–5. Bibcode:2009Natur.459..442T. doi:10.1038/nature07845. PMC 2743854Freely accessible. PMID 19404258.

94. Shukla VK, Doyon Y, Miller JC, et al. (May 2009). "Precise genome modification in the crop species Zea mays using zinc-finger nucleases". Nature. 459 (7245): 437–41. Bibcode:2009Natur.459..437S. doi:10.1038/nature07992. PMID 19404259.

95. Grizot S, Smith J, Daboussi F, et al. (September 2009). "Efficient targeting of a SCID gene by an engineered single-chain homing endonuclease". Nucleic Acids Res. 37 (16): 5405–19. doi:10.1093/nar/gkp548. PMC 2760784Freely accessible. PMID 19584299.

96. Gao H, Smith J, Yang M, et al. (January 2010). "Heritable targeted mutagenesis in maize using a designed endonuclease". Plant J. 61 (1): 176–87. doi:10.1111/j.1365-313X.2009.04041.x. PMID 19811621.

97. Christian M, Cermak T, Doyle EL, et al. (July 2010). "TAL Effector Nucleases Create Targeted DNA Double-strand Breaks". Genetics. 186 (2): 757–61. doi:10.1534/genetics.110.120717. PMC 2942870Freely accessible. PMID 20660643.

98. Li T, Huang S, Jiang WZ, et al. (August 2010). "TAL nucleases (TALNs): hybrid proteins composed of TAL effectors and FokI DNA-cleavage domain". Nucleic Acids Res. 39 (1): 359–72. doi:10.1093/nar/gkq704. PMC 3017587Freely accessible. PMID 20699274.

99. Ekker, S.C. (2008). "Zinc finger-based knockout punches for zebrafish genes". Zebrafish. 5 (2): 1121–3. doi:10.1089/zeb.2008.9988. PMC 2849655Freely accessible. PMID 18554175.

100. Geurts AM, Cost GJ, Freyvert Y, et al. (July 2009). "Knockout rats via embryo microinjection of zinc-finger nucleases". Science. 325 (5939): 433. Bibcode:2009Sci...325..433G. doi:10.1126/science.1172447. PMC 2831805Freely accessible. PMID 19628861.

101. Chen, I; Dubnau, D (2004). "DNA uptake during bacterial transformation". Nat. Rev. Microbiol. 2 (3): 241–9. doi:10.1038/nrmicro844. PMID 15083159.

102. Head, Graham; Hull, Roger H; Tzotzos, George T. (2009). Genetically Modified Plants: Assessing Safety and Managing Risk. London: Academic Pr. p. 244. ISBN 0-12-374106-8.

103. Gelvin, S. B. (2003). "Agrobacterium-Mediated Plant Transformation: The Biology behind the "Gene-Jockeying" Tool". Microbiology and Molecular Biology

Reviews. 67 (1): 16–37, table of contents. doi:10.1128/MMBR.67.1.16-37.2003. PMC 150518Freely accessible. PMID 12626681.

104. Darbani, Behrooz; Farajnia, Safar; Toorchi, Mahmoud; Zakerbostanabad, Saeed; Noeparvar, Shahin; Stewart, Jr., C. Neal (2010). "DNA-Delivery Methods to Produce Transgenic Plants". Science Alert.

105. Hohn, Barbara; Levy, Avraham A; Puchta, Holger (2001). "Elimination of selection markers from transgenic plants". Current Opinion in Biotechnology. 12 (2): 139–43. doi:10.1016/S0958-1669(00)00188-9. PMID 11287227.

106. Esvelt, KM.; Wang, HH. (2013). "Genome-scale engineering for systems and synthetic biology". Mol Syst Biol. 9: 641. doi:10.1038/msb.2012.66. PMC 3564264Freely accessible. PMID 23340847.

107. Tan, WS.; Carlson, DF.; Walton, MW.; Fahrenkrug, SC.; Hackett, PB. (2012). "Precision editing of large animal genomes". Adv Genet. Advances in Genetics. 80: 37–97. doi:10.1016/B978-0-12-404742-6.00002-8. ISBN 9780124047426. PMC 3683964Freely accessible. PMID 23084873.

108. "Natural Genetic Engineering and Natural Genome Editing". Annals of the New York Academy of Sciences. 1178. 2009.

109. Avise, John C. (2004). The hope, hype & reality of genetic engineering: remarkable stories from agriculture, industry, medicine, and the environment. Oxford University Press US. p. 22. ISBN 978-0-19-516950-8.

110. "Engineering algae to make complex anti-cancer 'designer' drug". PhysOrg. 10 December 2012. Retrieved 15 April 2013.

111. =Roque, AC; Lowe, CR; Taipa, MA. (2004). "Antibodies and genetically engineered related molecules: production and purification". Biotechnol Proress. 20 (3): 639–54. doi:10.1021/bp030070k. PMID 15176864.

112. Rodriguez, Luis L.; Grubman, Marvin J. (2009). "Foot and mouth disease virus vaccines". Vaccine. 27: D90–4. doi:10.1016/j.vaccine.2009.08.039. PMID 19837296.

113. "Background: Cloned and Genetically Modified Animals". Center for Genetics and Society. 14 April 2005.

114. "Knockout Mice". Nation Human Genome Research Institute. 2009.

115. "GM pigs best bet for organ transplant". Medical News Today. 21 September 2003.

116. Fischer, Alain; Hacein-Bey-Abina, Salima; Cavazzana-Calvo, Marina (2010). "20 years of gene therapy for SCID". Nature Immunology. 11 (6): 457–60. doi:10.1038/ni0610-457. PMID 20485269.

117. Ledford, Heidi (2011). "Cell therapy fights leukaemia". Nature. doi:10.1038/news.2011.472.

118. Lewitt, Peter A; Rezai, Ali R; Leehey, Maureen A; Ojemann, Steven G; Flaherty, Alice W; Eskandar, Emad N; Kostyk, Sandra K; Thomas, Karen; Sarkar, Atom; Siddiqui, Mustafa S; Tatter, Stephen B; Schwalb, Jason M; Poston, Kathleen L; Henderson, Jaimie M; Kurlan, Roger M; Richard, Irene H; Van Meter, Lori;

Sapan, Christine V; During, Matthew J; Kaplitt, Michael G; Feigin, Andrew (2011). "AAV2-GAD gene therapy for advanced Parkinson's disease: A double-blind, sham-surgery controlled, randomised trial". The Lancet Neurology. 10 (4): 309–19. doi:10.1016/S1474-4422(11)70039-4. PMID 21419704.

119. Gallagher, James. (2 November 2012) BBC News – Gene therapy: Glybera approved by European Commission. Bbc.co.uk. Retrieved on 15 December 2012.

120. Richards, Sabrina. "Gene Therapy Arrives in Europe". The Scientist. Retrieved 16 November 2012.

121. "1990 The Declaration of Inuyama". 5 August 2001. Archived from the original on 5 August 2001.

122. Smith KR, Chan S, Harris J. Human germline genetic modification: scientific and bioethical perspectives. Arch Med Res. 2012 Oct;43(7):491-513. doi: 10.1016/j.arcmed.2012.09.003. PMID 23072719

123. Wade, Nicholas (19 March 2015). "Scientists Seek Ban on Method of Editing the Human Genome". New York Times. Retrieved 20 March 2015.

124. Pollack, Andrew (3 March 2015). "A Powerful New Way to Edit DNA". New York Times. Retrieved 20 March 2015.

125. Baltimore, David; Berg, Paul; Botchan, Dana; Charo, R. Alta; Church, George; Corn, Jacob E.; Daley, George Q.; Doudna, Jennifer A.; Fenner, Marsha; Greely, Henry T.; Jinek, Martin; Martin, G. Steven; Penhoet, Edward; Puck, Jennifer; Sternberg, Samuel H.; Weissman, Jonathan S.; Yamamoto, Keith R. (19 March 2015). "A prudent path forward for genomic engineering and germline gene modification". Science. 348: 36–8. doi:10.1126/science.aab1028. PMC 4394183Freely accessible. PMID 25791083. Retrieved 20 March 2015.

126. Lanphier, Edward; Urnov, Fyodor; Haecker, Sarah Ehlen; Werner, Michael; Smolenski, Joanna (26 March 2015). "Don't edit the human germ line". Nature. 519: 410–411. doi:10.1038/519410a. PMID 25810189. Retrieved 20 March 2015.

127. Kolata, Gina (23 April 2015). "Chinese Scientists Edit Genes of Human Embryos, Raising Concerns". New York Times. Retrieved 24 April 2015.

128. Liang, Puping; et al. (18 April 2015). "CRISPR/Cas9-mediated gene editing in human tripronuclear zygotes". Protein & Cell. 6: 363–72. doi:10.1007/s13238-015-0153-5. PMC 4417674Freely accessible. PMID 25894090. Retrieved 24 April 2015.

129. Wade, Nicholas (3 December 2015). "Scientists Place Moratorium on Edits to Human Genome That Could Be Inherited". New York Times. Retrieved 3 December 2015.

130. Bergeson, Emilie R. (1997). "The Ethics of Gene Therapy".

131. Hanna, Kathi E. "Genetic Enhancement". National Human Genome Research Institute.

132. "Applications of Genetic Engineering". Microbiologyprocedure. Retrieved 9 July 2010.

133. "Biotech: What are transgenic organisms?". Easyscience. 2002. Retrieved 9 July 2010.

134. Savage, Neil (1 August 2007). "Making Gasoline from Bacteria: A biotech startup wants to coax fuels from engineered microbes". Technology Review. Retrieved 16 July 2015.

135. Summers, Rebecca (24 April 2013) Bacteria churn out first ever petrol-like biofuel New Scientist, Retrieved 27 April 2013

136. "Applications of Some Genetically Engineered Bacteria". Retrieved 9 July 2010.

137. Sanderson, Katherine (24 February 2012) New Portable Kit Detects Arsenic In Wells Chemical and Engineering News, Retrieved 23 January 2013

138. Reece, Jane B.; Urry, Lisa A.; Cain, Michael L.; Wasserman, Steven A.; Minorsky, Peter V.; Jackson, Robert B. (2011). Campbell Biology Ninth Edition. San Francisco: Pearson Benjamin Cummings. p. 421. ISBN 0-321-55823-5.

139. "New virus-built battery could power cars, electronic devices". Web.mit.edu. 2 April 2009. Retrieved 17 July 2010.

140. "Hidden Ingredient In New, Greener Battery: A Virus". Npr.org. Retrieved 17 July 2010.

141. "Researchers Synchronize Blinking 'Genetic Clocks' -- Genetically Engineered Bacteria That Keep Track of Time". ScienceDaily. 24 January 2010.

142. Suszkiw, Jan (November 1999). "Tifton, Georgia: A Peanut Pest Showdown". Agricultural Research magazine. Retrieved 23 November 2008.

143. Magaña-Gómez, JA; de la Barca, A.M. (2009). "Risk assessment of genetically modified crops for nutrition and health". Nutr. Rev. 67 (1): 1–16. doi:10.1111/j.1753-4887.2008.00130.x. PMID 19146501.

144. Islam, Aparna (2008). "Fungus Resistant Transgenic Plants: Strategies, Progress and Lessons Learnt". Plant Tissue Culture and Biotechnology. 16 (2): 117–38. doi:10.3329/ptcb.v16i2.1113.

145. "Disease resistant crops". GMO Compass.

146. Demont, M; Tollens, E (2004). "First impact of biotechnology in the EU: Bt maize adoption in Spain". Annals of Applied Biology. 145 (2): 197–207. doi:10.1111/j.1744-7348.2004.tb00376.x.

147. Whitman, Deborah B. (2000). "Genetically Modified Foods: Harmful or Helpful?".

148. Young, Emma (2003). "GM cows to please cheese-makers". New Scientist.

149. Rapeseed (canola) has been genetically engineered to modify its oil content with a gene encoding a "12:0 thioesterase" (TE) enzyme from the California bay plant (Umbellularia californica) to increase medium length fatty acids, see: Geo-pie.cornell.edu

150. Bomgardner Melody M (2012). "Replacing Trans Fat: New crops from Dow Chemical and DuPont target food makers looking for stable, heart-healthy oils". Chemical and Engineering News. 90 (11): 30–32.

151. Marvier, Michelle (2008). "Pharmaceutical crops in California, benefits and risks. A review". Agronomy for Sustainable Development. 28 (1): 1–9. doi:10.1051/agro:2007050.

152. "FDA Approves First Human Biologic Produced by GE Animals". US Food and Drug Administration.

153. Rebêlo, Paulo (15 July 2004). "GM cow milk 'could provide treatment for blood disease'". SciDev.

154. Pollack, Andrew (19 November 2015). "Genetically Engineered Salmon Approved for Consumption". The New York Times. Retrieved 21 April 2016.

155. "Giant GM salmon on the way". BBC News. 11 April 2000.

156. Chivian, Eric; Bernstein, Aaron (2008). Sustaining Life. Oxford University Press, Inc. ISBN 978-0-19-517509-7.

157. Carrington, Damien (13 June 2012) GM crops good for environment, study finds The Guardian. Retrieved 16 June 2012

158. Yang, Y.T.; Chen, B. (2016). "Governing GMOs in the USA: science, law and public health". Journal of the Science of Food and Agriculture. 96: 1851–1855. doi:10.1002/jsfa.7523. PMID 26536836. It is therefore not surprising that efforts to require labeling and to ban GMOs have been a growing political issue in the USA (citing Domingo and Bordonaba, 2011).

159. Bashshur, Ramona (February 2013). "FDA and Regulation of GMOs". American Bar Association. Retrieved February 24, 2016.

160. Sifferlin, Alexandra (October 3, 2015). "Over Half of E.U. Countries Are Opting Out of GMOs". Time.

161. Lynch, Diahanna; Vogel, David (April 5, 2001). "The Regulation of GMOs in Europe and the United States: A Case-Study of Contemporary European Regulatory Politics". Council on Foreign Relations. Retrieved February 24, 2016.

162. "Can GM crops harm the environment?". National Environment Research Council (NERC).

163. Angulo, E.; Cooke, B. (2002). "First synthesize new viruses then regulate their release? The case of the wild rabbit". Molecular Ecology. 11: 2703–9. doi:10.1046/j.1365-294X.2002.01635.x. PMID 12453252.

164. Adams; et al. (2 August 2002). "The Case for Genetic Engineering of Native and Landscape Trees against Introduced Pests and Diseases". Conservation Biology. 16: 874–879. doi:10.1046/j.1523-1739.2002.00523.x. Retrieved 16 May 2016.

165. Thomas; et al. (25 September 2013). "Ecology: Gene tweaking for conservation". Nature. 501: 485–6. doi:10.1038/501485a. PMID 24073449. Retrieved 16 May 2016.

166. Pasko, Jessica M. (2007-03-04). "Bio-artists bridge gap between arts, sciences: Use of living organisms is attracting attention and controversy". msnbc.

167. Jackson, Joab (6 December 2005). "Genetically Modified Bacteria Produce Living Photographs". National Geographic News.

168. Phys.Org website. 4 April 2005 "Plant gene replacement results in the world's only blue rose".

169. Katsumoto, Yukihisa; Fukuchi-Mizutani, Masako; Fukui, Yuko; Brugliera, Filippa; Holton, Timothy A.; Karan, Mirko; Nakamura, Noriko; Yonekura-Sakakibara, Keiko; Togami, Junichi; Pigeaire, Alix; Tao, Guo-Qing; Nehra, Narender S.; Lu, Chin-Yi; Dyson, Barry K.; Tsuda, Shinzo; Ashikari, Toshihiko; Kusumi, Takaaki; Mason, John G.; Tanaka, Yoshikazu (2007). "Engineering of the Rose Flavonoid Biosynthetic Pathway Successfully Generated Blue-Hued Flowers Accumulating Delphinidin". Plant and Cell Physiology. 48 (11): 1589–600. doi:10.1093/pcp/pcm131. PMID 17925311.

170. Published PCT Application WO2000049150 "Chimeric Gene Constructs for Generation of Fluorescent Transgenic Ornamental Fish." National University of Singapore

171. Stewart, C. Neal (2006). "Go with the glow: Fluorescent proteins to light transgenic organisms". Trends in Biotechnology. 24 (4): 155–62. doi:10.1016/j.tibtech.2006.02.002. PMID 16488034.

172. WHO (1987): Principles for the Safety Assessment of Food Additives and Contaminants in Food, Environmental Health Criteria 70. World Health Organization, Geneva

173. WHO (1991): Strategies for assessing the safety of foods produced by biotechnology, Report of a Joint FAO/WHO Consultation. World Health Organization, Geneva ISBN 9789241561457. PDF download library.health.go.ug/download/file/fid/790 here]

174. WHO (1993): Health aspects of marker genes in genetically modified plants, Report of a WHO Workshop. World Health Organization, Geneva

175. WHO (1995): Application of the principle of substantial equivalence to the safety evaluation of foods or food components from plants derived by modern biotechnology, Report of a WHO Workshop. World Health Organization, Geneva

176. Johnston C, Martin B, Fichant G, Polard P, Claverys JP (2014). "Bacterial transformation: distribution, shared mechanisms and divergent control". Nat. Rev. Microbiol. 12 (3): 181–96. doi:10.1038/nrmicro3199. PMID 24509783.

177. Alberts, Bruce; et al. (2002). Molecular Biology of the Cell. New York: Garland Science. p. G:35. ISBN 978-0-8153-4072-0.

178. Griffith, F. 1928. The significance of pneumococcal types. J. Hyg. 27: 113–159, .

179. Lederberg, Joshua (1994). The Transformation of Genetics by DNA: An Anniversary Celebration of AVERY, MACLEOD and MCCARTY(1944) in Anecdotal, Historical and Critical Commentaries on Genetics. The Rockfeller University, New York, New York 10021-6399. PMID 8150273.

180. Mandel, Morton; Higa, Akiko (1970). "Calcium-dependent bacteriophage DNA infection". Journal of Molecular Biology. 53 (1): 159–162. doi:10.1016/0022-2836(70)90051-3. PMID 4922220.

181. Cohen, Stanley N.; Chang, Annie; Hsu, Leslie (1972). "Nonchromosomal Antibiotic Resistance in Bacteria: Genetic Transformation of Escherichia coli by R-Factor DNA". Proceedings of the National Academy of Sciences. 69 (8): 2110–4. doi:10.1073/pnas.69.8.2110. PMC 426879Freely accessible. PMID 4559594.

182. Hanahan, D. (1983). "Studies on transformation of Escherichia coli with plasmids". Journal of Molecular Biology. 166 (4): 557–580. doi:10.1016/S0022-2836(83)80284-8. PMID 6345791.

183. Wirth R, Friesenegger A, Fiedler S (March 1989). "Transformation of various species of gram-negative bacteria belonging to 11 different genera by electroporation". Mol. Gen. Genet. 216 (1): 175–7. doi:10.1007/BF00332248. PMID 2659971.

184. Palmiter, Richard; Ralph L. Brinster; Robert E. Hammer; Myrna E. Trumbauer; Michael G. Rosenfeld; Neal C. Birnberg; Ronald M. Evans (1982). "Dramatic growth of mice that develop from eggs microinjected with metallothionein−growth hormone fusion genes". Nature. 300 (5893): 611–5. doi:10.1038/300611a0. PMID 6958982.

185. Nester, Eugene. "Agrobacterium: The Natural Genetic Engineer (100 Years Later)". Retrieved 14 January 2011.

186. Zambryski, P.; Joos, H.; Genetello, C.; Leemans, J.; Montagu, M. V.; Schell, J. (1983). "Ti plasmid vector for the introduction of DNA into plant cells without alteration of their normal regeneration capacity". The EMBO Journal. 2 (12): 2143–2150. PMC 555426Freely accessible. PMID 16453482.

187. Peters, Pamela. "Transforming Plants - Basic Genetic Engineering Techniques". Retrieved 28 January 2010.

188. "Biologists invent gun for shooting cells with DNA" (PDF). Cornell Chronicle. 14 May 1987. p. 3.

189. Sanford JC, et al. (1987). "Delivery of substances into cells and tissues using a particle bombardment process". Journal of Particulate Science and Technology. 5: 27–37. doi:10.1080/02726358708904533.

190. Klein TM, Wolf ED, Wu R, Sanford JC (7 May 1987). "High-velocity microprojectiles for delivering nucleic acids into living cells". Nature. 327 (6117): 70–73. doi:10.1038/327070a0.

191. Michod RE, Bernstein H, Nedelcu AM (2008). "Adaptive value of sex in microbial pathogens". Infect. Genet. Evol. 8 (3): 267–85. doi:10.1016/j.meegid.2008.01.002. PMID 18295550.

192. Seitz P, Blokesch M (2013). "Cues and regulatory pathways involved in natural competence and transformation in pathogenic and environmental Gram-negative bacteria". FEMS Microbiol. Rev. 37 (3): 336–63. doi:10.1111/j.1574-6976.2012.00353.x. PMID 22928673.

193. Johnsborg O, Eldholm V, Håvarstein LS (2007). "Natural genetic transformation: prevalence, mechanisms and function". Res. Microbiol. 158 (10): 767–78. doi:10.1016/j.resmic.2007.09.004. PMID 17997281.

194. Cite error: The named reference Alberts was invoked but never defined (see the help page).

195. Chen I, Dubnau D (March 2004). "DNA uptake during bacterial transformation". Nat. Rev. Microbiol. 2 (3): 241–9. doi:10.1038/nrmicro844. PMID 15083159.

196. Lacks, S.; Greenberg, B.; Neuberger, M. (1974). "Role of a Deoxyribonuclease in the Genetic Transformation of Diplococcus pneumoniae". Proceedings of the National Academy of Sciences of the United States of America. 71 (6): 2305–2309. doi:10.1073/pnas.71.6.2305. PMC 388441Freely accessible. PMID 4152205.

197. Long, C. D.; Tobiason, D. M.; Lazio, M. P.; Kline, K. A.; Seifert, H. S. (2003). "Low-Level Pilin Expression Allows for Substantial DNA Transformation Competence in Neisseria gonorrhoeae". Infection and immunity. 71 (11): 6279–6291. doi:10.1128/iai.71.11.6279-6291.2003. PMC 219589Freely accessible. PMID 14573647.

198. Sisco, K. L.; Smith, H. O. (1979). "Sequence-specific DNA uptake in Haemophilus transformation". Proceedings of the National Academy of Sciences of the United States of America. 76 (2): 972–976. doi:10.1073/pnas.76.2.972. PMC 383110Freely accessible. PMID 311478.

199. Johnsborg O, Eldholm V, Håvarstein LS (December 2007). "Natural genetic transformation: prevalence, mechanisms and function". Res. Microbiol. 158 (10): 767–78. doi:10.1016/j.resmic.2007.09.004. PMID 17997281.

200. Solomon JM, Grossman AD (April 1996). "Who's competent and when: regulation of natural genetic competence in bacteria". Trends Genet. 12 (4): 150–5. doi:10.1016/0168-9525(96)10014-7. PMID 8901420.

201. Saito Y, Taguchi H, Akamatsu T (March 2006). "Fate of transforming bacterial genome following incorporation into competent cells of Bacillus subtilis: a continuous length of incorporated DNA". J. Biosci. Bioeng. 101 (3): 257–62. doi:10.1263/jbb.101.257. PMID 16716928.

202. Saito Y, Taguchi H, Akamatsu T (April 2006). "DNA taken into Bacillus subtilis competent cells by lysed-protoplast transformation is not ssDNA but dsDNA". J. Biosci. Bioeng. 101 (4): 334–9. doi:10.1263/jbb.101.334. PMID 16716942.

203. Akamatsu T, Taguchi H (April 2001). "Incorporation of the whole chromosomal DNA in protoplast lysates into competent cells of Bacillus subtilis". Biosci. Biotechnol. Biochem. 65 (4): 823–9. doi:10.1271/bbb.65.823. PMID 11388459.

204. Goodgal SH, Herriott RM (July 1961). "Studies on transformations of Hemophilus influenzae. I. Competence". J. Gen. Physiol. 44 (6): 1201–27. doi:10.1085/jgp.44.6.1201. PMC 2195138Freely accessible. PMID 13707010.

205. Aspiras MB, Ellen RP, Cvitkovitch DG (September 2004). "ComX activity of Streptococcus mutans growing in biofilms". FEMS Microbiol. Lett. 238 (1): 167–74. doi:10.1016/j.femsle.2004.07.032. PMID 15336418.

206. Anagnostopoulos C, Spizizen J (May 1961). "Requirements for transformation in Bacillus subtilis". J. Bacteriol. 81 (5): 741–6. PMC 279084Freely accessible. PMID 16561900.

207. Keen, Eric C.; Bliskovsky, Valery V.; Malagon, Francisco; Baker, James D.; Prince, Jeffrey S.; Klaus, James S.; Adhya, Sankar L.; Groisman, Eduardo A. (2017). "Novel "Superspreader" Bacteriophages Promote Horizontal Gene Transfer by Transformation". mBio. 8 (1): e02115–16. doi:10.1128/mBio.02115-16. ISSN 2150-7511.

208. Claverys JP, Prudhomme M, Martin B (2006). "Induction of competence regulons as a general response to stress in gram-positive bacteria". Annu. Rev. Microbiol. 60: 451–75. doi:10.1146/annurev.micro.60.080805.142139. PMID 16771651.

209. Michod RE, Wojciechowski MF, Hoelzer MA (January 1988). "DNA repair and the evolution of transformation in the bacterium Bacillus subtilis". Genetics. 118 (1): 31–9. PMC 1203263Freely accessible. PMID 8608929.

210. Dorer MS, Fero J, Salama NR (2010). Blanke, Steven R, ed. "DNA damage triggers genetic exchange in Helicobacter pylori". PLoS Pathog. 6 (7): e1001026. doi:10.1371/journal.ppat.1001026. PMC 2912397Freely accessible. PMID 20686662.

211. Charpentier X, Kay E, Schneider D, Shuman HA (March 2011). "Antibiotics and UV radiation induce competence for natural transformation in Legionella pneumophila". J. Bacteriol. 193 (5): 1114–21. doi:10.1128/JB.01146-10. PMC 3067580Freely accessible. PMID 21169481.

212. Albertini S, Chételat AA, Miller B, et al. (July 1995). "Genotoxicity of 17 gyrase- and four mammalian topoisomerase II-poisons in prokaryotic and eukaryotic test systems". Mutagenesis. 10 (4): 343–51. doi:10.1093/mutage/10.4.343. PMID 7476271.

213. Washburn RS, Gottesman ME (January 2011). "Transcription termination maintains chromosome integrity". Proc. Natl. Acad. Sci. U.S.A. 108 (2): 792–7. doi:10.1073/pnas.1009564108. PMC 3021005Freely accessible. PMID 21183718.

214. Sakano K, Oikawa S, Hasegawa K, Kawanishi S (November 2001). "Hydroxyurea induces site-specific DNA damage via formation of hydrogen peroxide and nitric oxide". Jpn. J. Cancer Res. 92 (11): 1166–74. doi:10.1111/j.1349-7006.2001.tb02136.x. PMID 11714440.

215. Bernstein H, Bernstein C, Michod RE (2012). "Chapter 1". In Sakura Kimura and Sora Shimizu. DNA repair as the primary adaptive function of sex in bacteria and eukaryotes. DNA Repair: New Research. Nova Sci. Publ., Hauppauge, N.Y. pp. 1–49. ISBN 978-1-62100-808-8.

216. Michod RE, Bernstein H, Nedelcu AM (May 2008). "Adaptive value of sex in microbial pathogens". Infect. Genet. Evol. 8 (3): 267–85. doi:10.1016/j.meegid.2008.01.002. PMID 18295550. http://www.hummingbirds.arizona.edu/Faculty/Michod/Downloads/IGE%20review%20sex.pdf

217. Donahue RA, Bloom FR (July 1998). "Large-volume transformation with high-throughput efficiency chemically competent cells" (PDF). Focus. 20 (2): 54–56. OCLC 12352630.

218. Srivastava, Sheela (2013). Genetics of Bacteria (PDF). India: Springer-Verlag. doi:10.1007/978-81-322-1090-0. ISBN 978-81-322-1089-4.

219. Kawai, Shigeyuki; Hashimoto, Wataru; Murata, Kousaku (1 November 2010). "Transformation of Saccharomyces cerevisiae and other fungi: methods and possible underlying mechanism.". Bioengineered Bugs. 1 (6): 395–403. doi:10.4161/bbug.1.6.13257.

220. Hinnen, A; Hicks, JB; Fink, GR (April 1978). "Transformation of yeast.". Proceedings of the National Academy of Sciences of the United States of America. 75 (4): 1929–33. doi:10.1073/pnas.75.4.1929. PMID 347451.

221. Ito, H; Fukuda, Y; Murata, K; Kimura, A (January 1983). "Transformation of intact yeast cells treated with alkali cations". Journal of Bacteriology. 153 (1): 163–8. PMC 217353Freely accessible. PMID 6336730.

222. Gietz, RD; Woods, RA (2002). "Transformation of yeast by lithium acetate/single-stranded carrier DNA/polyethylene glycol method". Methods in enzymology. Methods in Enzymology. 350: 87–96. doi:10.1016/S0076-6879(02)50957-5. ISBN 9780121822538. PMID 12073338.

223. Gietz, RD; Schiestl, RH; Willems, AR; Woods, RA (Apr 15, 1995). "Studies on the transformation of intact yeast cells by the LiAc/SS-DNA/PEG procedure". Yeast. 11 (4): 355–60. doi:10.1002/yea.320110408. PMID 7785336.

224. Schiestl, Robert H.; Manivasakam, P.; Woods, Robin A.; Gietzt, R.Daniel (1 August 1993). "Introducing DNA into Yeast by Transformation". Methods. 5 (2): 79–85. doi:10.1006/meth.1993.1011.

225. Spencer, F.; Ketner, G.; Connelly, C.; Hieter, P. (1 August 1993). "Targeted Recombination-Based Cloning and Manipulation of Large DNA Segments in Yeast". Methods. 5 (2): 161–175. doi:10.1006/meth.1993.1021.

226. Costanzo, MC; Fox, TD (November 1988). "Transformation of yeast by agitation with glass beads". Genetics. 120 (3): 667–70. PMC 1203545Freely accessible. PMID 3066683.

227. Dohmen, R. J.; Strasser, A. W.; Höner, C. B.; Hollenberg, C. P. (1991). "An efficient transformation procedure enabling long-term storage of competent cells of various yeast genera". Yeast. 7 (7): 691–2. doi:10.1002/yea.320070704. PMID 1776359.

228. Hayama, Y; Fukuda, Y; Kawai, S; Hashimoto, W; Murata, K (2002). "Extremely simple, rapid and highly efficient transformation method for the yeast Saccharomyces cerevisiae using glutathione and early log phase cells". Journal of bioscience and bioengineering. 94 (2): 166–71. PMID 16233287.

229. V.Singh and D.K.Jain (2014). "Applications of recombinant DNA". ISC BIOLOGY. Nageen Prakashan. p. 840.

230. Bacterial Transformation

231. Inoue, H.; Nojima, H.; Okayama, H. (1990). "High efficiency transformation of Escherichia coli with plasmids". Gene. 96 (1): 23–28. doi:10.1016/0378-1119(90)90336-P. PMID 2265755.

232. Donahue RA, Bloom FR (September 1998). "Transformation efficiency of E. coli electroporated with large plasmid DNA" (PDF). Focus. 20 (3): 77–78. Archived from the original on September 3, 2011.

233. H.C. Bimboim; J. Doly (3 August 1979). "A rapid alkaline extraction procedure for screening recombinant plasmid DNA" (PDF). Nucleic Acids Research. 7 (6): 1513–1523. doi:10.1093/nar/7.6.1513. Archived (PDF) from the original on 21 December 2016. Retrieved 21 December 2016.

CHAPTER 4

Selection Methods in Plant Breeding Based on Mode of Reproduction

Chapter Contents

INTRODUCTION

Plant breeders use different methods depending on the mode of reproduction of crops, which include:

- Self-fertilization, where pollen from a plant will fertilise reproductive cells or ovules of the same plant
- Cross-pollination, where pollen from one plant can only fertilize a different plant
- Asexual propagation (e.g. runners from strawberry plants) where the new plant is genetically identical to its parent
- Apomixis (self-cloning), where seeds are produced asexually and the new plant is genetically identical to its parent

The mode of reproduction of a crop determines its genetic composition, which, in turn, is the deciding factor to develop suitable breeding and selection methods. Knowledge of mode of reproduction is also essential for its artificial manipulation to breed improved types. Only those breeding and selection methods are suitable for a crop which does not interfere with its natural state or ensure the maintenance of such a state. It is due to such reasons that imposition of self-fertilization on cross-pollinating crops leads to drastic reduction in their performance.

For teaching purpose, plant breeding is presented as four categories: Line breeding (autogamous crops), population breeding (allogamous crops), hybrid breeding (mostly allogamous crops, some autogamous crops), clone breeding (vegetatively propagated crops).

SELF FERTILIZING CROPS (AUTOGAMOUS CROPS)

Certain restrictions caused the mechanisms for self-fertilization (partial and full self-fertilization) to develop in a number of plant species. Some of the reasons why a self-fertilizing method of reproduction is so effective are the efficacy of reproduction, as well as decreasing genetic variation and thus the fixation of highly adapted genotypes. Almost no inbreeding depression occurs in self-fertilizing plants because the mode of reproduction allows natural selection to take place in wild populations of such plants.

Critical steps in the improvement of self-fertilizing crops are the choice of parents and the identification of the best plants in segregating generations. The breeder should also have definite goals with the choice of parents. Self-fertilizing are easier to maintain, but this could lead to misuse of seed.

Some of the agronomy important, self-fertilizing crops include wheat, rice, barley, dry beans, soy beans, peanuts, tomatoes, etc.

MASS SELECTION

This method of selection depends mainly on selection of plants according to their phenotype and performance. The seed from selected plants are bulked for the next generation. This method is used to improve the overall population by positive or negative mass selection. Mass selection is only applied to a limited degree in self-fertilizing plants and is an effective method for the improvement of land races. This method of selection will only be effective for highly heritable traits. One shortage of mass selection are the large influence that the environment has on the development, phenotype and performance of single plants. This can also be an advantage in that varieties can be selected for local performance.

Stratified mass selection for ear size over 22 cycles has drastically altered plant phenotype in the maize population Zacatecas 58. Plants in the C22 cycle were 50 cm taller, had twice the leaf area index, reached anthesis 7 days later and had a 30% higher harvest index than C0 (Table 1) Differences in growth were detected early in ontogeny. The root growth of C22 exceeded that of C0 and the ratio of shoot dry mass to root dry mass was reduced by nearly 12%, from 8.0±0.2 to 7.1±0.1 (Table 2). Analysis of yield components revealed that C22 was superior to C0 in grain weight, number of rows per ear, number of grains per row, and total yield per unit area (Table 3). Because the two genotypes were phenologically different, planting density optima are probably different for each population.

SELECTION OF CROSS-POLLINATED CROPS

Plant species where normal mode of seed set is through a high degree of cross-pollination have characteristic reproductive features and population structure. Existence of self-sterility, self-incompatibility, imperfect flowers, and mechanical obstructions make the plant dependent upon foreign pollen for normal seed set. Each plant receives a blend of pollen from a large number of individuals each having different genotypes. Such populations are characterized by a high degree of heterozygosity with tremendous free and potential genetic variation, which is maintained in a steady state by free gene flow among individuals within the populations.

In the development of hybrid varieties, the aim is to identify the most productive heterozygote from the population, which then is produced with the exclusion of other members of the population.

Mass Selection

It is the simplest, easiest and oldest method of selection where individual plants are selected based on their phenotypic performance, and bulk seed selection proved to be quite effective in maize improvement at the initial stages but its efficacy especially for improvement of yield, soon came under severe criticism that culminated in the refinement of the method of mass selection. The selection after pollination does not provide any control over the pollen parent as result of which effective selection is limited only to female parents. The heritability estimates are reduced by half, since only parents are used to harvest seed whereas the pollen source is not known after the cross pollination has taken place.

Recurrent Selection

This type of selection is a refined version of the mass selection procedure and differs as follows:

- Visually selected individuals out of the base population undergo progeny testing
- Individuals selected on basis of the progeny test data are crossed with each other in every possible way to produce seed to form the new base population.

Half-Sib Selection with Progeny Testing

Selections are made based on progeny test performance instead of phenotypic appearance of the parental plants. Seed from selected half-sibs, which have been pollinated by random pollen from the population (meaning that only the female parent is known and selected, hence the term "half-sib") is grown in unreplicated progeny rows for the purpose of selection. A part of the seed is planted to determine the yielding ability, or breeding value, for any character of each plant. The seed from the most productive rows or remnant seed from the outstanding half-sibs is bulked to complete one cycle of selection.

Full-Sib Selection with Progeny Testing

A number of full-sib families, each produced by making crosses between the two plants from the base population are evaluated in replicated trials. A part of each full-sib family is saved for recombination. Based on evaluation the remnant seed of selected full-sib families is used to recombine the best families.

BREEDING OF ASEXUALLY PROPAGATED CROPS

Asexual reproduction covers all those modes of multiplication of plants where normal gamete formation and fertilization does not take place making these distinctly different from normal seed production crops. In the absence of sexual reproduction, the genetic composition of plant material being multiplied remains essentially the same as its source plant.

Clones of mother plants can be made with the exact genetic composition of the mother plant. Superior plants are selected and propagated vegetatively; the vegetative propagated offspring are used to develop stable varieties without any deterioration due to segregation of gene combinations. This unique characteristic of asexual reproduction helped to develop a number of cultivars of fruits and vegetables including grapes, apples, pears and peaches.

Improving Asexual Plant Material through Selection

The selection in these crops is restricted to the material introduced from other sources, such as field plantations. The improvement of asexually propagated plants through induced mutations has distinct advantages and limitations. Any vegetative propagule can be treated with mutagens and even a single desirable mutant or a part of a mutated propagule (chimera) can be multiplied as an improved type of the original variety.

Selection of Asexual Plants

Selection, in the case of asexual plants, can be defined as the selection of the best performing plant and the vegetative propagation thereof. Because plants are not totally genetically stable, it can be expected that deviations would occur through the years. Selection is thus an ongoing process where deviants are selected or removed from the selection program. The main purpose of selection is to better the quality and yield of forthcoming plantations. Different approaches can be followed in the selection process of asexual plants, such as mass selection and clone selection from clone blocks.

In mass selection there are some factors that must be considered when selecting plants in a mother block, e.g. vineyard. Time of selection is a big factor, because you have to select when most of the characteristics of the plant are clearly showing. With asexual perennials the best time is just before harvest. For the best results the selected plant must be evaluated during the next season, when growth-abnormalities, leave disfigurations and virus symptoms are best visualized. Mass selection is done annually on the same plant for a minimum of three years. A plant that does not conform to the requirements in any given year of the selection cycle is discarded from the program.

NEW CLONE DEVELOPMENT

The development and registration of new clones take place by means of local clone selection in old plantations, as well as the importation of high quality clones from abroad, for local evaluation.

A clone is the vegetative offspring of one specific mother plant; it does not show any genetic, morphologic or physiologic deviations from the mother plant. Evaluation takes place with the different selected clones after selection.

PLANT PROPAGATION

Plant propagation is the process of creating new plants from a variety of sources: seeds, cuttings, bulbs and other plant parts. Plant propagation can also refer to the artificial or natural dispersal of plants.

Gentian seedlings in a plant nursery.

Sexual Propagation (Seed)

One way to propagate an avocado seed

Seeds and spores can be used for reproduction (through e.g. sowing). Seeds are typically produced from sexual reproduction within a species, because genetic recombination has occurred. A plant grown from seeds may have different characteristics from its parents. Some species produce seeds that require special conditions to germinate, such as cold treatment. The seeds of many Australian plants and plants from southern Africa and the American west require smoke or fire to germinate. Some plant species, including many trees do not produce seeds until they reach maturity, which may take many years. Seeds can be difficult to acquire and some plants do not produce seed at all. Some plants (like certainF1/F2 hybrids and GMO plants) may produce seed, but not fertile seed. In certain cases (like with GMO's), this is done to prevent the accidental spreading of these plants (which are generally non-native crops), for example by birds and other animals.

Asexual Propagation

Plants have a number of mechanisms for asexual or vegetative reproduction. Some of these have been taken advantage of by horticulturists and gardeners to multiply or clone plants rapidly. People also use methods that plants do not use, such as tissue culture and grafting. Plants are produced using material from a single parent and as such there is no exchange of genetic material, therefore vegetative propagation methods almost always produce plants that are identical to the parent. Vegetative reproduction uses plants parts such as roots, stems and leaves. In some plants seeds can be produced without fertilization and the seeds contain only the genetic material of the parent plant. Therefore, propagation via asexual seeds or apomixis is asexual reproduction but not vegetative propagation.

Softwood stemcuttings rooting in a controlled environment.

Techniques for vegetative propagation include:

- Air or ground layering
- Division
- Grafting and bud grafting, widely used in fruit tree propagation
- Micropropagation
- Stolons or runners
- Storage organs such as bulbs, corms, tubers and rhizomes
- Striking or cuttings
- Twin-scaling
- Offsets

Layering

Layering is a means of plant propagation in which a portion of an aerial stem grows roots while still attached to the parent plant and then detaches as an independent plant. Layering has evolved as a common means of vegetative propagation of numerous species in natural environments. Layering is also utilized by horticulturists to propagate desirable plants.

Natural layering typically occurs when a branch touches the ground, whereupon it produces adventitious roots. At a later stage the connection with the parent plant is severed and a new plant is produced as a result.

The horticultural layering process typically involves wounding the target region to expose the inner stem and optionally applying rooting compounds. In ground layering or simple layering, the stem is bent down and the target region buried in the soil. This is done in plant nurseries in imitation of natural layering by many plants such as brambles which bow over and touch the tip on the ground, at which point it grows roots and, when separated, can continue as a separate plant. In either case, the rooting process may take from several weeks to a year.

Layering is more complicated than taking cuttings, but has the advantage that the propagated portion continues to receive water and nutrients from the parent plant while it is forming roots. This is important for plants that form roots slowly, or for propagating large pieces. Layering is used quite frequently in the propagation of bonsai; it is also used as a technique for both creating new roots and improving existing roots.

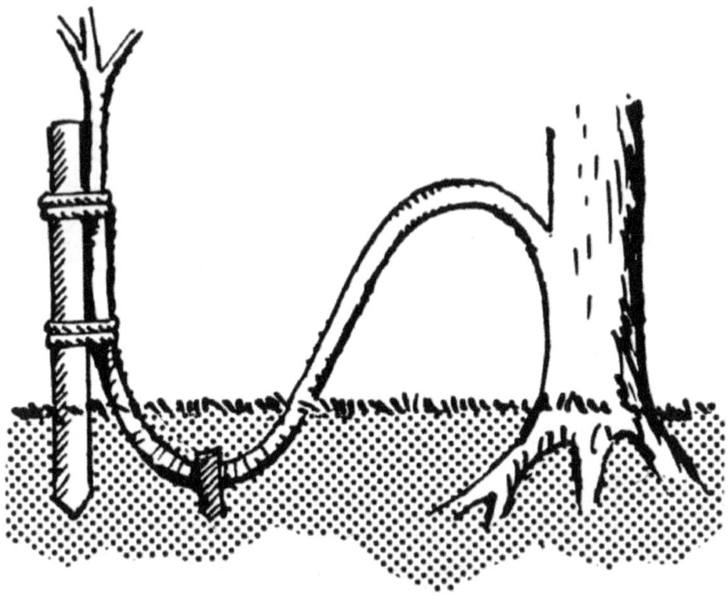

Layering.

Method

A low-growing stem is bent down to touch a hole dug in the ground, then pinned in place using something shaped like a clothes hanger hook and covered over with soil. However, a few inches of leafy growth must remain above the ground for the bent stem to grow into a new plant. Removing a section of skin from the lower-facing stem part before burying may help the rooting process. If using rooting hormone, the stem should be cut just beneath a node. The resultant notch should be wedged open with a toothpick or similar piece of wood and the hormone applied before burying.

The buried stem part then grows new roots which take hold in the soil while the above parts grow as individual new plants. Once the end of the stem has grown long enough the process can be repeated, creating the appearance of a row of plants linked by humped, intermittently buried stems. Better results can be achieved when the top of the plant is closer to the vertical.

Once the process is completed, the buried section should be kept well-watered until roots begin to form. The new individual plant may require one to two years before it is strong enough to survive on its own. When it is, the original stem should be cut where it enters the ground, thereby separating the two plants.

Plant Types and Benefits

As layering does not involve sexual reproduction, new plants are effectively clones of the original plant and will exhibit the same characteristics. This includes flower, fruit and foliage. Plant selection usually involves plants with a flexible stem.

Simple layering can be more attractive when managing a cascading or spreading plant. These plants tend to propagate in this manner anyway, and potting a new limb will give extra plants without having to sow new seed.

Simple layering can also help when a plant has overgrown its pot and is drooping over the side. The long stem is layered into another pot until it roots, thus bringing it back to soil level.

Ground Layering

Ground layering or mound layering is the typical propagation technique for the popular Malling-Merton series of clonal apple rootstocks, in which the original plants are set in the ground with the stem nearly horizontal, which forces side buds to grow upward . After these are started, the original stem is buried up to some distance from the tip. At the end of the growing season, the side branches will have rooted, and can be separated while the plant is dormant. Some of these will be used for grafting rootstocks, and some can be reused in the nursery for the next growing season's crop.

Ground layering is used in the formation of visible surface roots, known as "nebari", on bonsai trees.

Air Layering

In air layering (or marcotting), the target region is wounded, or a strip of bark is removed and then encased in a moisture-retaining medium such as sphagnum moss, or cloth which is further surrounded in a moisture barrier such as plastic film. Add soil and the scraped skin. Rooting hormone is often applied to encourage the wounded region to grow roots. When sufficient roots have grown from the wound, the stem is removed from the parent plant and planted. It can take the layer from a few weeks to one or more growing seasons to produce sufficient roots; this is largely dependent on the plant species and the vigor of the parent plant.

Air layering of *Limonium dendroides.*

Division (horticulture)

Division, in horticulture and gardening, is a method of asexual plant propagation, where the plant (usually an herbaceous perennial) is broken up into two or more parts. Both the root and crown of each part is kept intact. The technique is of ancient origin, and has long been used to propagate bulbs such as garlic and saffron. Division is mainly practiced by gardeners and very small nurseries, as most commercial plant propagation is now done through plant tissue culture.

Division is one of the three main methods used by gardeners to increase stocks of plants (the other two are seed-sowing and cuttings). Division is usually applied to mature perennial plants, but may also be used for shrubs with suckering roots, such as gaultheria, kerria and sarcococca. Annual and biennial plants do not lend themselves to this procedure, as their lifespan is too short.

Most perennials are best divided and replanted every few years to keep them healthy. They may also be divided in order to produce new plants. Those with woody crowns or fleshy roots need to be cut apart, while others can be prized apart using garden forks or hand forks. Each separate section must have both shoots and roots. Division can take place at almost any time of the year, but the best seasons are Autumn and Spring.

Grafting

Grafting or graftage is a horticultural technique whereby tissues of plants are joined so as to continue their growth together. The upper part of the combined plant is called the scion while the lower part is called the rootstock. The success of this joining requires that the vascular tissue grow together and such joining is called inosculation. The technique is most commonly used in asexual propagation of commercially grown plants for the horticultural and agricultural trades.

In most cases, one plant is selected for its roots and this is called the stock or rootstock. The other plant is selected for its stems, leaves, flowers, or fruits and is called the scion or cion. The scion contains the desired genes to be duplicated in future production by the stock/scion plant.

In stem grafting, a common grafting method, a shoot of a selected, desired plant cultivar is grafted onto the stock of another type. In another common form called bud grafting, a dormant side bud is grafted onto the stem of another stock plant, and when it has inosculated successfully, it is encouraged to grow by pruning off the stem of the stock plant just above the newly grafted bud.

For successful grafting to take place, the vascular cambium tissues of the stock and scion plants must be placed in contact with each other. Both tissues must be kept alive until the graft has "taken", usually a period of a few weeks. Successful grafting only requires that a vascular connection take place between the grafted tissues. Joints formed by grafting are not as strong as naturally formed joints, so a physical weak point often still occurs at the graft because only the newly formed tissues inosculate with each other. The existing structural tissue (or wood) of the stock plant does not fuse.

Cherry tree, consolidated "V" graft.

Tape has been used to bind the rootstock and scion at the graft and tar of the scion from desiccation.

Advantages

Graft particular to plum Cherry. The scion is the largest in the plant, due to the imperfect union of the two. It can be seen on the enlarged trunk: this accumulation of starch is an indication of imperfection.

- Precocity: The ability to induce fruitfulness without the need for completing the juvenile phase. Juvenility is the natural state through which a seedling plant must pass before it can become reproductive. In most fruiting trees, juvenility may last between 5 and 9 years, but in some tropical fruits e.g. Mangosteen, juvenility may be prolonged for up to 15 years. Grafting of mature scions onto rootstocks can result in fruiting in as little as two years.
- Dwarfing: To induce dwarfing or cold tolerance or other characteristics to the scion. Most apple trees in modern orchards are grafted on to dwarf or semi-dwarf trees planted at high density. They provide more fruit per unit of land, higher quality fruit, and reduce the danger of accidents by harvest crews working on ladders. Care must be taken when planting dwarf or semi-dwarf trees. If such a tree is planted with the graft below the soil, then the scion portion can also grow roots and the tree will still grow to its standard size.
- Ease of propagation: Because the scion is difficult to propagate vegetatively by other means, such as by cuttings. In this case, cuttings of an easily rooted plant are used to provide a rootstock. In some cases, the scion may be easily propagated, but grafting may still be used because it is commercially the most cost-effective way of raising a particular type of plant.
- Hybrid breeding: To speed maturity of hybrids in fruit tree breeding programs. Hybrid seedlings may take ten or more years to flower and fruit on their own roots. Grafting can reduce the time to flowering and shorten the breeding program.
- Hardiness: Because the scion has weak roots or the roots of the stock plants are tolerant of difficult conditions. e.g. many showy Western Australian plants are sensitive to dieback on heavy soils, common in urban gardens, and are grafted onto hardier eastern Australian relatives. Grevilleas and eucalypts are examples.
- Sturdiness: To provide a strong, tall trunk for certain ornamental shrubs and trees. In these cases, a graft is made at a desired height on a stock plant with a strong stem. This is used to raise 'standard' roses, which are rose bushes on a high stem, and it is also used for some ornamental trees, such as certain weeping cherries.
- Disease/Pest Resistance: In areas where soil-borne pests or pathogens would prevent the successful planting of the desired cultivar, the use of pest/disease tolerant rootstocks allow the production from the cultivar that would be otherwise unsuccessful. A major example is the use of rootstocks in combating Phylloxera.
- Pollen source: To provide pollenizers. For example, in tightly planted or badly planned apple orchards of a single variety, limbs of crab apple may be grafted at regularly spaced intervals onto trees down rows, say every fourth tree. This takes care of pollen needs at blossom time, yet does not confuse pickers who might otherwise mix varieties while harvesting, as the mature crab apples are so distinct from other apple varieties.
- Repair: To repair damage to the trunk of a tree that would prohibit nutrient flow, such as stripping of the bark by rodents that completely girdles the trunk. In this case a bridge graft may be used to connect tissues receiving flow from the roots to tissues above the damage that have been severed from the flow.

Where a water-shoot, basal shoot or sapling of the same species is growing nearby, any of these can be grafted to the area above the damage by a method called inarch grafting. These alternatives to scions must be of the correct length to span the gap of the wound.

- Changing cultivars: To change the cultivar in a fruit orchard to a more profitable cultivar, called *top working*. It may be faster to graft a new cultivar onto existing limbs of established trees than to replant an entire orchard.
- Maintain consistency: Apples are notorious for their genetic variability, even differing in multiple characteristics, such as, size, color, and flavor, of fruits located on the same tree. In the commercial farming industry, consistency is maintained by grafting a scion with desired fruit traits onto a hardy stock.

An example of approach grafting by Axel Erlandson.

- Curiosities
- A practice sometimes carried out by gardeners is to graft related potatoes and tomatoes so that both are produced on the same plant, one above ground and one underground.
- Cacti of widely different forms are sometimes grafted on to each other.
- Multiple cultivars of fruits such as apples are sometimes grafted on a single tree. This so-called "family tree" provides more fruit variety for small spaces such as a suburban backyard, and also takes care of the need for pollenizers.

The drawback is that the gardener must be sufficiently trained to prune them correctly, or one strong variety will usually "take over."

- Ornamental and functional, tree shaping uses grafting techniques to join separate trees or parts of the same tree to itself. Furniture, hearts, entry archways are examples. Axel Erlandson was a prolific tree shaper who grew over 75 mature specimens.

Techniques

Approach

T budding.

Approach grafting or inarching is used to join together plants that are otherwise difficult to join. The plants are grown close together, and then joined so that each plant has roots below and growth above the point of union. Both scion and stock retain their respective parents that may or may not be removed after joining. Also used in pleaching. The graft can be successfully accomplished any time of year.

Budding

Bud grafting uses a bud instead of a twig. Grafting roses is the most common example of bud grafting. In this method a bud is removed from the parent plant, and the base of the bud is inserted beneath the bark of the stem of the stock plant from which the rest of the shoot has been cut. Any extra bud that starts growing out from the stem of the stock plant is removed because that would bear the flower of the unwanted original kind. Examples: roses and peaches. Budding is the easiest technique used.

Budwood is a stick with several buds on it that can be cut out and used for bud grafting. It is a common method of propagation for citrus trees.

Cleft

In cleft grafting a small cut is made in the stalk and then the pointed end of the scion is inserted in the stalk. The most common form of grafting is cleft grafting. This is best done in the spring and is useful for joining a thin scion about 1 cm (0.39 in) diameter to a thicker branch or stock. It is best if the latter is 2–7 cm (0.79–2.76 in) in diameter and has 3–5 buds. The branch or stock should be split carefully down the middle to form a cleft about 3 cm (1.2 in) deep. If it is a branch that is not vertical then the cleft should be cut horizontally. The end of the scion should be cut cleanly to a long shallow wedge, preferably with a single cut for each wedge surface, and not whittled. A third cut may be made across the end of the wedge to make it straight across.

Successful cleft graft after 2 years' growth.

Same graft after 4 years' growth. Graft is now final, rootstock completely overtaken

Slide the wedge into the cleft so that it is at the edge of the stock and the centre of the wedge faces are against the cambium layer between the bark and the wood. It is preferable if a second scion is inserted in a similar way into the other side of the cleft. This helps to seal off the cleft. Tape around the top of the stock to hold the scion/s in place and cover with grafting wax or sealing compound. This stops the cambium layers from drying out and also prevents the ingress of water into the cleft.

Whip

Successful whip graft.

Successful whip graft needing additional pruning the following season.

In whip grafting the scion and the stock are cut slanting and then joined. The grafted point is then bound with tape and covered with soft wax to prevent dehydration and germs. Also known as the whip and tongue graft, this is considered the most difficult to master but has the highest rate of success as it offers the most cambium contact between the two species. It is the most common graft used in top-dressing commercial fruit trees. It is generally used with stock less than $\frac{1}{2}$ in (1.3 cm) diameter, with the ideal diameter closer to $\frac{3}{8}$ in (0.95 cm) and the scion should be of roughly the same diameter as the stock.

The stock is cut through on one side only at a shallow angle with a sharp knife. (If the stock is a branch and not the main trunk of the rootstock then the cut surface should face outward from the centre of the tree.) The scion is similarly sliced through at an equal angle starting just below a bud, so that the bud is at the top of the cut and on the other side than the cut face.

A notch is cut downwards into the sliced face of the stock and a similar cut upwards into the face of the scion cut. These act as the tongues and it requires some skill to make the cuts so that the scion and the stock marry up neatly. The joint is then taped around and treated with tree-sealing compound or grafting wax.

The elongated "Z" shape adds strength, removing the need for a companion rod in the first season.

Stub

Stub grafting is a technique that requires less stock than cleft grafting, and retains the shape of a tree. Also scions are generally of 6-8 buds in this process.

Successful stub graft, healed.

An incision is made into the branch 1 cm (0.39 in) long, then the scion is wedged and forced into the branch. The scion should be at an angle of at most 35° to the parent tree so that the crotch remains strong. The graft is covered with grafting compound.

After the graft has taken, the branch is removed and treated a few centimeters above the graft, to be fully removed when the graft is strong.

Four Flap (or Banana)

The four-flap graft (also called banana graft) is commonly used for pecans, and first became popular with this species in Oklahoma in 1975. It is heralded for maximum cambium overlap, but is a complex graft. It requires similarly sized diameters for the rootstock and graftwood. The bark of the rootstock is sliced and peeled back in four flaps, and the hardwood is removed, looking somewhat like a peeled banana. It is a difficult graft to learn.

Awl

Awl grafting takes the least resources and the least time. It is best done by an experienced grafter, as it is possible to accidentally drive the tool too far into the stock, reducing the scion's chance of survival. Awl grafting can be done by using a screwdriver to make a slit in the bark, not penetrating the cambium layer completely. Then inset the wedged scion into the incision.

Veneer

Veneer grafting, or inlay grafting, is a method used for stocks larger than three centimeters in diameter. The scion is recommended to be about as thick as a pencil. Clefts are made of the same size as the scion on the side of the branch, not on top. The scion end is shaped as a wedge, inserted, and wrapped with tape to the scaffolding branches to give it more strength.

Natural Grafting

Tree branches and more often roots of the same species will sometimes naturally graft; this is called inosculation. When roots make physical contact with each other they often grow together. A group of trees can share water and mineral nutrients via root grafts, which may be advantageous to weaker trees, and may also form a larger rootmass as an adaptation to promote fire resistance and regeneration as exemplified by the California black oak (*Quercus kelloggii*). Albino redwoods use root grafting as a form of plant parasitism of normal redwoods.

Possible deliberate grafts on a Sessile Oak in Ayrshire, Scotland.

A Husband and Wife tree - Natural grafting in blackthorn Prunus spinosa.

A problem with root grafts is that they allow transmission of certain pathogens, such as Dutch elm disease. Inosculation also sometimes occurs where two stems on the same tree, shrub or vine make contact with each other. This is common in plants such as strawberries and potato.

Graft Chimera

Occasionally, a so-called "graft hybrid" or more accurately graft chimera can occur where the tissues of the stock continue to grow within the scion. Such a plant can produce flowers and foliage typical of both plants as well as shoots intermediate between the two. The best-known example this is probably *+Laburnocytisus* 'Adamii', a graft hybrid between laburnum and broom, which originated in a nursery near Paris, France in 1825. This small tree bears yellow flowers typical of *Laburnum anagyroides*, purple flowers typical of *Chamaecytisus purpureus* and curious coppery-pink flowers that show characteristics of both "parents". Many species of cactus can also produce graft chimeras under the right conditions although they are often created unintentionally and such results are often hard to replicate.

Scientific Uses

Grafting has been important in flowering research. Leaves or shoots from plants induced to flower can be grafted onto uninduced plants and transmit a floral stimulus that induces them to flower.

The transmission of plant viruses has been studied using grafting. Virus indexing involves grafting a symptomless plant that is suspected of carrying a virus onto an indicator plant that is very susceptible to the virus.

Grafting can transfer chloroplasts (specialised DNA in plants that can conduct photosynthesis), mitochondrial DNA and the entire cell nucleus containing the genome to potentially make a new species making grafting a form of natural genetic engineering.

White Spruce

White spruce can be graffted with consistent success by using 8 cm to 10 cm scions of current growth on thrifty 4- to 5-year-old rootstock (Nienstaedt and Teich 1972). Before greenhouse grafting, rootstocks should be potted in late spring, allowed to make seasonal growth, then subjected to a period of chilling outdoors, or for about 8 weeks in a cool room at 2 °C (Nienstaedt 1966).

A method of grafting white spruce of seed-bearing age during the time of seed harvest in the fall was developed by Nienstaedt et al. (1958). Scions of white spruce of 2 ages of wood from 30- to 60-year-old trees were collected in the fall and grafted by 3 methods on potted stock to which different day-length treatments had been applied prior to grafting. The grafted stock were given long-day and natural-day treatments. Survival was 70% to 100% and showed effects of rootstock and post-grafting treatments in only a few cases. Photoperiod and temperature treatments after grafting, however, had considerable effect on scion activity and total growth. The best post-grafting treatment was 4 weeks of long-day treatment followed by 2 weeks of short-day treatment, then 8 weeks of chilling, and finally long-day treatment.

Since grafts of white spruce put on relatively little growth in the 2 years after grafting, techniques for accelerating the early growth were studied by Greenwood (1988) and others. The cultural regimes used to promote one additional growth cycle in one year involve manipulation of day length and the use of cold storage to satisfy chilling requirements. Greenwood took dormant potted grafts into the greenhouse in early January then gradually raised the temperature during the course of a week until the minimum temperature rose to 15 °C. Photoperiod was increased to 18 hours using incandescent lighting. In this technique, grafts are grown until elongation has been completed, normally by mid-March. Soluble 10-52-10 fertilizer is applied at both ends of the growth cycle and 20-20-20 during the cycle, with irrigation as needed. When growth elongation is complete, day length is reduced to 8 hours using a blackout curtain. Budset follows, and the grafts are held in the greenhouse until mid-May. Grafts are then moved into a cooler at 4 °C for 1000 hours, after which they are moved to a shade frame where they grow normally, with applications of fertilizer and irrigation as in the first cycle. Grafts are moved into cold frames or unheated greenhouse in September until January. Flower induction treatments are begun on grafts that have reached a minimum length of 1.0 m. Repotting from an initial pot size of 4.5litre to 16litre containers with a 2:1:1 soil mix of peat moss, loam, and aggregate.

In one of the first accelerated growth experiments, white spruce grafts made in January and February that would normally elongate shortly after grafting, set bud, and remain in that condition until the following spring, were refrigerated for 500, 1000, or 1500 hours beginning in mid-July, and a non-refrigerated control was held in the nursery. After completion of the cold treatment, the grafts were moved into the greenhouse with an 18-hour photoperiod until late October. Height increment was significantly (P 0.01) influenced by cold treatment. Best results were given by the 1000-hour treatment.

The refrigeration (cold treatment) phase was subsequently shown to be effective when applied 2 months earlier with proper handling and use of blackout curtains, which allows the second growth cycle to be completed in time to satisfy dormancy requirements before January (Greenwood et al. 1988).

Herbaceous Grafting

Grafting is often done for non -woody and vegetable plants (tomato, cucumber, eggplant and watermelon). Tomato grafting is very popular in Asia and Europe, and is gaining popularity in the United States. The main advantage of grafting is for disease-resistant rootstocks. Researchers in Japan developed automated processes using grafting robots as early as 1987. Plastic tubing can be used to prevent desiccation and support the healing at the graft/scion interface.

Budding

Budding is a form of asexual reproduction in which a new organism develops from an outgrowth or bud due to cell division at one particular site. The new organism remains attached as it grows, separating from the parent organism only when it is mature, leaving behind scar tissue. Since the reproduction is asexual, the newly created organism is a clone and is genetically identical to the parent organism.

Organisms such as hydra use regenerative cells for reproduction in the process of budding. In hydra, a bud develops as an outgrowth due to repeated cell division at one specific site. These buds develop into tiny individuals and, when fully mature, detach from the parent body and become new independent individuals.

Internal budding or endodyogeny is a process of asexual reproduction, favored by parasites such as *Toxoplasma gondii*. It involves an unusual process in which two daughter cells are produced inside a mother cell, which is then consumed by the offspring prior to their separation.

Endopolygeny is the division into several organisms at once by internal budding.

Cellular Reproduction

Some cells divide asymmetrically by budding, for example *Saccharomyces cerevisiae*, the yeast species used in baking and brewing. This process results in a 'mother' cell and a smaller 'daughter' cell. Cryo-electron tomography recently revealed that mitochondria in cells divide by budding.

Animal Reproduction

Hydra with two buds.

In some multicellular animals (metazoans), offspring may develop as outgrowths of the mother. Animals that reproduce by budding include corals, some sponges, some acoel flatworms (e.g., *Convolutriloba*), and echinoderm larvae.

Colony Division

Colonies of some bee species have also exhibited budding behavior, such as *Apis dorsata*. Although budding behavior is rare in this bee species, it has been observed when a group of workers leave the natal nest and construct a new nest usually near the natal one.

Virology

In virology, *budding* is a form of viral shedding by which enveloped viruses acquire their external envelope from the host cell membrane, which bulges outwards and encloses the virion.

Micropropagation

Micropropagation is the practice of rapidly multiplying stock plant material to produce a large number of progeny plants, using modern plant tissue culture methods.

Micropropagation is used to multiply noble plants such as those that have been genetically modified or bred through conventional plant breeding methods. It is also used to provide a sufficient number of plantlets for planting from a stock plant which does not produce seeds, or does not respond well to vegetative reproduction.

Cornell University botanist Frederick Campion Steward discovered and pioneered micropropagation and plant tissue culture in the late 1950s and early 1960s.

A rose plant that began as cells grown in a tissue culture.

Stages of Micropropagation

Establishment

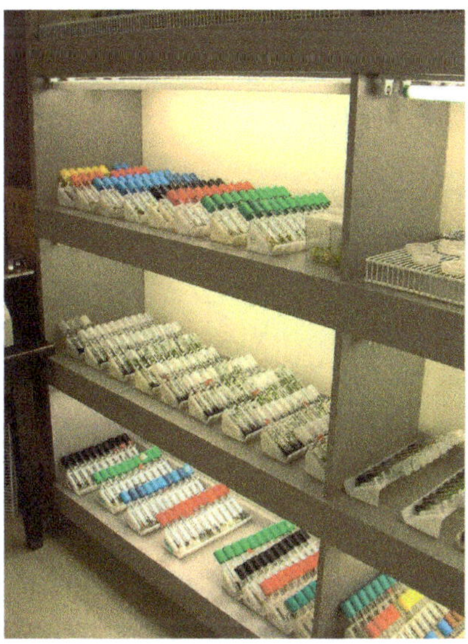

In vitro culture of plants in a controlled, sterile environment.

Micropropagation begins with the selection of plant material to be propagated. The plant tissues are removed from an intact plant in a sterile condition. Clean stock materials that are free of viruses and fungi are important in the production of the healthiest plants. Once the plant material is chosen for culture, the collection of explant(s) begins and is dependent on the type of tissue to be used; including stem tips, anthers, petals, pollen and others plant tissues. The explant material is then surface sterilized, usually in multiple courses of bleach and alcohol washes, and finally rinsed in sterilized water. This small portion of plant tissue, sometimes only a single cell, is placed on a growth medium, typically containing sucrose as an energy source and one or more plant growth regulators (plant hormones). Usually the medium is thickened with agar to create a gel which supports the explant during growth. Some plants are easily grown on simple media, but others require more complicated media for successful growth; the plant tissue grows and differentiates into new tissues depending on the medium. For example, media containing cytokinin are used to create branched shoots from plant buds.

Multiplication

Multiplication is the taking of tissue samples produced during the first stage and increasing their number. Following the successful introduction and growth of plant tissue, the establishment stage is followed by multiplication. Through repeated cycles of this process, a single explant sample may be increased from one to hundreds and thousands of plants. Depending on the type of tissue grown, multiplication can involve different methods and media. If the plant material grown is callus tissue, it can be placed in a blender and cut into smaller pieces and recultured on the same type of culture medium to grow more callus tissue. If the tissue is grown as small plants called plantlets, hormones are often added that cause the plantlets to produce many small offshoots. After the formation of multiple shoots, these shoots are transferred to rooting medium with a high auxin\cytokinin ratio. After the development of roots, plantlets can be used for hardening.

Pretransplant

Banana plantlets transferred to soil (with vermicompost) from plant media. This process is done for acclimatization of plantlets to the soil as they were previously grown in plant media. After growing for some days the plantlets are transferred to the field.

This stage involves treating the plantlets/shoots produced to encourage root growth and "hardening." It is performed *in vitro*, or in a sterile "test tube" environment.

"Hardening" refers to the preparation of the plants for a natural growth environment. Until this stage, the plantlets have been grown in "ideal" conditions, designed to encourage rapid growth. Due to the controlled nature of their maturation, the plantlets often do not have fully functional dermal coverings. This causes them to be highly susceptible to disease and inefficient in their use of water and energy. In vitro conditions are high in humidity, and plants grown under these conditions often do not form a working cuticle and stomata that keep the plant from drying out. When taken out of culture, the plantlets need time to adjust to more natural environmental conditions. Hardening typically involves slowly weaning the plantlets from a high-humidity, low light, warm environment to what would be considered a normal growth environment for the species in question.

Transfer from Culture

Plant tissue cultures being grown at a USDA seed bank, the National Center for Genetic Resources Preservation.

In the final stage of plant micropropagation, the plantlets are removed from the plant media and transferred to soil or (more commonly) potting compost for continued growth by conventional methods.

This stage is often combined with the "pretransplant" stage.

Methods of Micro-propagation

Meristem Culture

In Meristem culture the Meristem and a few subtending leaf primordial are placed into a suitable growing media. An elongated rooted platelet is produced after some weeks, and is transferred to the soil when it has attained a considerable height. A disease free plant can be produced by this method. Experimental result also suggest that this technique can be successfully utilized for rapid multiplication of various herbaceous Plants.

Callus Culture

A callus is mass of undifferentiated parenchymatous cells. When a living plant tissue is placed in an artificial growing medium with other conditions favorable, callus is formed. The growth of callus varies with the homogenous levels of auxin and Cyotkininn and can be manipulated by endogenous supply of these growth regulators in the culture medium. The callus growth and its organogenesis or embryogenesis can be referred into three different stages.

- Stage I: Rapid production of callus after placing the explants in culture medium
- Stage II: The callus is transferred to other medium containing growth regulators for the induction of adventitious organs.
- Stage III: The new plantlet is then exposed gradually to the environmental condition.

Suspension Culture

A cell suspension culture refers to cells and or groups of cells dispersed and growing in an aerated liquid culture medium (Street, 1997, Thorpe1981) is placed in a liquid medium and shaken vigorously and balanced dose of hormones. Suezawa et al. (1988) reported Cyotkininn induced adventitious buds in kiwi fruit in a suspension culture sub- culture for about a week.

Embryo Culture

In embryo culture, the embryo is excised and placed into a culture medium with proper nutrient in aseptic condition. To obtain a quick and optimum growth into plantlets, it is transferred to soil. It is particularly important for the production of interspecific and intergeneric hybrids and to overcome the embryo abortion.

Protoplast Culture

In protoplast culture, the plant cell can be isolated with the help of wall degrading enzymes and growth in a suitable culture medium in a controlled condition for regeneration of plantlets. Under suitable conditions the protoplast develops a cell wall followed by an increase in cell division and differentiation and grows into a new plant. The protoplast are fist culture in liquid medium at 25 to 28 C with a light intensity of 100 to 500 lux or in dark and after undergoing substantial cell division, they are transferred into solid medium congenial or morphogenesis in many horticultural crops response well to protoplast culture.

Advantages

Micropropagation has a number of advantages over traditional plant propagation techniques:

- The main advantage of micropropagation is the production of many plants that are clones of each other.

- Micropropagation can be used to produce disease-free plants.
- It can have an extraordinarily high fecundity rate, producing thousands of propagules while conventional techniques might only produce a fraction of this number.
- It is the only viable method of regenerating genetically modified cells or cells after protoplast fusion.
- It is useful in multiplying plants which produce seeds in uneconomical amounts, or when plants are sterile and do not produce viable seeds or when seed cannot be stored.
- Micropropagation often produces more robust plants, leading to accelerated growth compared to similar plants produced by conventional methods - like seeds or cuttings.
- Some plants with very small seeds, including most orchids, are most reliably grown from seed in sterile culture.
- A greater number of plants can be produced per square meter and the propagules can be stored longer and in a smaller area.

Disadvantages

Micropropagation is not always the perfect means of multiplying plants. Conditions that limits its use include:

- It is *very* expensive, and can have a labour cost of more than 70%.
- A monoculture is produced after micropropagation, leading to a lack of overall disease resilience, as all progeny plants may be vulnerable to the same infections.
- An infected plant sample can produce infected progeny. This is uncommon as the stock plants are carefully screened and vetted to prevent culturing plants infected with virus or fungus.
- Not all plants can be successfully tissue cultured, often because the proper medium for growth is not known or the plants produce secondary metabolic chemicals that stunt or kill the explant.
- Sometimes plants or cultivars do not come true to type after being tissue cultured. This is often dependent on the type of explant material utilized during the initiation phase or the result of the age of the cell or propagule line.
- Some plants are very difficult to disinfect of fungal organisms.

The major limitation in the use of micropropagation for many plants is the cost of production; for many plants the use of seeds, which are normally disease free and produced in good numbers, readily produce plants in good numbers at a lower cost. For this reason, many plant breeders do not utilize micropropagation because the cost is prohibitive. Other breeders use it to produce stock plants that are then used for seed multiplication.

Mechanisation of the process could reduce labour costs, but has proven difficult to achieve, despite active attempts to develop technological solutions.

Stolon

In biology, **stolons** (from Latin stolō "branch"), also known as runners, are horizontal connections between organisms. They may be part of the organism, or of its skeleton; typically, animal stolons are external skeletons.

Pingao (Ficinia spiralis) spreads by forming stolons in the sand.

Common Silverweed (Argentina anserina) picture showing red stolons.

In Botany

In botany, stolons are stems which grow at the soil surface or just below ground that form adventitious roots at the nodes, and new plants from the buds. Stolons are often called runners. Rhizomes, in contrast, are root-like stems that may either grow horizontally at the soil surface or in other orientations underground. Thus, not all horizontal stems are called stolons. Plants with stolons are called stoloniferous.

A stolon is a plant propagation strategy and the complex of individuals formed by a mother plant and all its clones produced from stolons form a single genetic individual, a genet.

Morphology

These stolons from the corm of a *Crocosmia* are stems that emerged from axillary buds at the nodes of the tunic leaves.

Stolons may or may not have long internodes. The leaves along the stolon are usually very small, but in a few cases such as *Stachys sylvatica* are normal in size.

Stolons arise from the base of the plant. In strawberries the base is above the soil surface; in many bulb-forming species and plants with rhizomes, the stolons remain underground and form shoots that rise to the surface at the ends or from the nodes. The nodes of the stolons produce roots, often all around the node and hormones produced by the roots cause the stolon to initiate shoots with normal leaves. Typically after the formation of the new plant the stolon dies away in a year or two, while rhizomes persist normally for many years or for the life of the plant, adding more length each year to the ends with active growth. The horizontal growth of stolons results from the interplay of different hormones produced at the growing point and hormones from the main plant, with some studies showing that stolon and rhizome growth are affected by the amount of shady light the plant receives with increased production and branching from plants exposed to mixed shade and sun, while plants in all day sun or all shade produce fewer stolons.

A number of plants have soil-level or above-ground rhizomes, including *Iris* species and many orchid species.

T. Holm (1929) restricted the term rhizome to a horizontal, usually subterranean, stem that produces roots from its lower surface and green leaves from its apex, developed directly from the plumule of the embryo. He recognized stolons as axillary, subterranean branches that do not bear green leaves but only membranaceous, scale-like ones.

Plants with Stolons

'Lipstick' hybrid strawberry (*Comarum palustre* × *Fragaria* × *ananassa*) along stolons.

In some *Cyperus* species the stolons end with the growth of tubers; the tubers are swollen stolons that form new plants.

Some species of crawling plants can also sprout adventitious roots, but are not considered stoloniferous: a stolon is sprouted from an existing stem and can produce a full individual. Examples of plants that extend through stolons include some species from the genera *Argentina* (silverweed), *Cynodon*, *Fragaria*, and *Pilosella* (Hawkweeds), *Zoysia japonica*, *Ranunculus repens*.

Other plants with stolons below the soil surface include many grasses, *Ajuga*, *Mentha*, and *Stachys*.

Lily-of-the-valley (*Convallaria majalis*) has rhizomes that grow stolon-like stems called stoloniferous rhizomes or leptomorph rhizomes. A number of plants have stoloniferous rhizomes including Asters. These stolon-like rhizomes are long and thin, with long internodes and indeterminate growth with lateral buds at the node, which mostly remain dormant.

In potatoes, the stolons start to grow within 10 days of plants emerging above ground, with tubers usually beginning to form on the end of the stolons. The tubers are modified stolons that hold food reserves, with a few buds that grow into stems. Since it is *not* a rhizome it does not generate roots, but the new stem growth that grows to the surface produces roots.

Hydrilla use stolons that produce tubers to spread themselves and to survive dry periods in aquatic habitats.

Erythronium, commonly called Trout Lily, have white stolons growing from the bulb. Most run horizontally, either underground or along the surface of the ground under leaf litter. A number of bulbous species produce stolons, such as *Erythronium propullans*. Flowering plants often produce no stolons.

Convolvulus arvensis is a weed species in agriculture that spreads by under ground stolons that produce rhizomes.

In studies on grass species, with plants that produce stolons or rhizomes and plants that produce both stolons and rhizomes, morphological and physiological differences were noticed. Stolons have longer internodes and function as means of seeking out light and are used for propagation of the plant, while rhizomes are used as storage organs for carbohydrates and the maintenance of meristem tissue to keep the parent plant alive from one year to the next.

In Mycology

In mycology, a stolon is defined as an occasionally septate hypha, which connects sporangiophores together. Root-like structures called rhizoids may appear on the stolon as well, anchoring the hyphae to the substrate. The stolon is commonly found in bread molds, and are seen as horizontally expanding across the mold.

Cutting

A plant **cutting** is a piece of a plant that is used in horticulture for vegetative (asexual) propagation. A piece of the stem or root of the source plant is placed in a suitable medium such as moist soil. If the conditions are suitable, the plant piece will begin to grow as a new plant independent of the parent, a process known as **striking**. A stem cutting produces new roots, and a root cutting produces new stems. Some plants can be grown from leaf pieces, called leaf cuttings, which produce both stems and roots. The scions used in grafting are also called cuttings.

A magnolia stem cutting has been coaxed to form new roots, and is now a complete plant.

Technique

Some plants form roots much more easily than others. Stem cuttings from woody plants are treated differently, depending on the maturity of the wood:

- Softwood cuttings come from stems that are rapidly expanding, with young leaves. In many species, such cuttings form roots relatively easily.
- Semi-hardwood cuttings come from stems that have completed elongation growth and have mature leaves.
- Hardwood cuttings come from fully matured stems, and are often propagated while dormant.

Softwood cuttings of elm (*Ulmus*) are kept under a water mist to prevent them from drying out while they form roots.

Most plant cuttings are stem pieces, and have no root system of their own, they are likely to die from dehydration if the proper conditions are not met. They require a moist medium, which, however, cannot be too wet lest the cutting rot. A number of media are used in this process, including but not limited to soil, perlite, vermiculite, coir, rock wool, expanded clay pellets, and even water given the right conditions. Most succulent cuttings can be left in open air until the cut surface dries, which may improve root formation when the cutting is later planted.

In temperate countries, stem cuttings may be taken of soft (green or semi-ripe) wood and hard wood which has specific differences in practice. Certain conditions lead to more favorable outcomes for cuttings; timing, size, location on the plant, and amount of foliage are all important. In temperate countries, stem cuttings of young wood need to be taken in spring from the upper branches, while of hardened wood need to be taken in winter from the lower branches. Common bounds on the length of stem cuttings are between 5–15 centimetres (2.0–5.9 in) for soft wood and between 20–25 centimetres (7.9–9.8 in) for hard wood. Soft wood cuttings do best when about two thirds of the foliage removed, while hard wood stem cuttings need complete foliage removal.

Cuttings from a variety of succulents.

The environment for cuttings is generally kept humid—often attained by placing the cuttings under a plastic sheet or in another confined space where the air can be kept moist—and partial shade to prevent the cutting from drying out. Cuttings in the medium are typically watered with a fine mist to avoid disturbing plants. Following the initial watering, the aim to keep the soil moist but not wet and waterlogged; the medium is allowed to almost dry out before misting again.

A rooting hormone may be administered to "encourage" growth and maturity in plants determined to be unlikely to grow. Though not essential, several compounds may be used to promote the formation of roots through the signaling activity of plant hormone auxins, and is helpful with especially hard plant species. Among the commonly used chemicals is indole-3-butyric acid (IBA) used as a powder, liquid solution or gel. This compound is applied either to the cut tip of the cutting or as a foliar spray. Rooting hormone can be manufactured naturally, such as soaking the yellow-tipped shoots of a weeping willow tree in water or to preparing a tea from the bark of a willow tree. Shoots or bark do better when soaked for 24 hours prior to using. Honey, though it does not contain any plant hormones, can also aid in rooting success through its antiseptic quality.

Types of Cuttings

Many vegetative parts of a plant can be used. The most common methods are:

- Stem cuttings, in which a piece of stem is part buried in the soil, including at least one leaf node. The cutting is able to produce new roots, usually at the node.
- Root cuttings, in which a section of root is buried just below the soil surface, and produces new shoots
- Scion cuttings are used in grafting.
- Leaf cuttings, in which a leaf is placed on moist soil. These have to develop both new stems and new roots. Some leaves will produce one plant at the base

of the leaf. In some species, multiple new plants can be produced at many places on one leaf, and these can be induced by cutting the leaf veins.

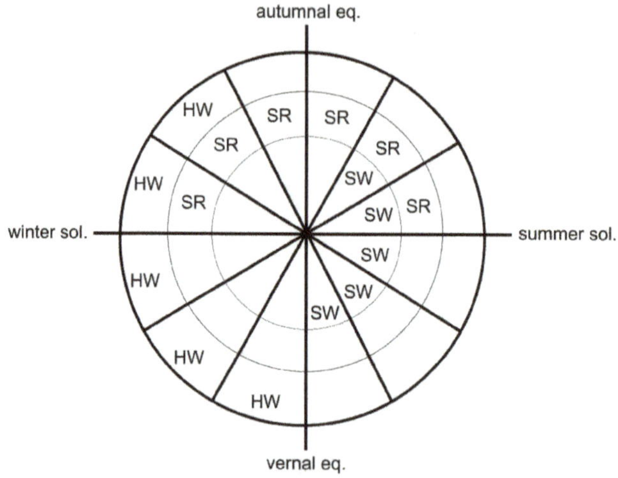

Scheme of appropriate type of stem cuttings according to season, based on several sources. Key: eq.:equinox, sol.: solstice, HW: hardwood, SR: semi-ripe, SW: softwood.

Although some species, such as willow, blackberry and pelargoniums can be grown simply by placing a cutting into moist ground, the majority of species require more attention. Most species require humid, warm, partially shaded conditions to strike, thus requiring the approach above to be followed. Particularly difficult species may need cool air above and warm soil. In addition, with many more difficult cuttings, one should use the type of cutting that has the most chance of success with that particular plant species.

Improving Results

Rose cuttings under plastic bottle greenhouse.

Softwood stemcuttings of *Buxus sempervirens*.

There are ways of improving the growth of stem cutting propagations. Intensifying light allows cuttings to root and sprout faster, though the heat thus generated could cause the propagation material distress. Azalea cuttings can be mildly heated in water to disinfect it from the fungus pathogen Rhizoctonia, and this could potentially be used for other plants.

Providing the Right Soil

Depending on the type of soil being used, several additives may need adding to create good soil for cuttings. These additions may include:

- chalk; to increase the pH-value of the soil; a pH of 6-6.5 is to be maintained
- organic substance/humus; to increase nutrient load; keep to a bare minimum though
- sand or gravel; to increase the soil's water permeability

For example, with plain potting soil, a third of the container should be filled with sand, to make suitable soil for cuttings.

Providing the Right Humidity

A white plastic greenhouse, used to keep the cuttings humid.

Although several options can be used here, usually semi-white plastic is used to cover the cuttings. The soil below and from the cuttings themselves is kept moist, and should be aerated once in a while to prevent formation of molds. A plastic bottle can be used as a small greenhouse to provide the right humidity level.

Twin-Scaling

Twin-scaling is a method of propagating plant bulbs that have a basal plate, such as:

- *Hippeastrum*, *Narcissus*, *Galanthus* and other members of the *Amaryllidaceae*;
- some members of the lily family *Liliaceae*;
- *Lachenalia*, *Veltheimia* and other members of the *Hyacinthaceae*.

Purpose

Twin-scaling is practiced by professional growers and skilled amateurs to increase bulbs that would naturally propagate very slowly, or to speed up the production of desirable cultivars. Using twin-scaling it is possible to multiply one bulb into 16 to 32 (or more) viable bulbs in a couple of years, whereas natural propagation might only lead to a doubling every two years or so. It is one of a number of propagation techniques (such as “scooping”, “scoring” and “chipping”) based on the fact that an accidentally damaged bulb will often regenerate by forming small bulblets or bulbils on the damaged surface. Commercial growers have obtained as many as 100 twin-scales from a single bulb.

Method

The dormant bulb which is to be twin-scaled has its surface sterilized by removing its dry tunic and carefully trimming off its roots and any dead tissue, while leaving a layer of sound basal plate intact, then dipping the clean bulb in dilute bleach (or another suitable disinfectant). The bulb is then sliced cleanly from top to bottom several times, creating 8 or 16 segments, depending on the size of the bulb. At this stage the segments are called "chips" (many growers are content with simply chipping a bulb into 4 or 8 and do not divide the bulb further).

True twin-scaling involves further subdivision of the chips to create pairs of scales, joined together by a small part of the basal plate. The twin-scales are then treated with fungicide before being mixed with moist, sterile Vermiculite, sealed in plastic bags and left in a fairly warm, dark location until new bulblets form. Some species may require alternate periods of warm and cool storage to initiate bulblet growth.

The tiny bulbs are planted into pots or trays of compost and allowed to grow on for a year or more, until they are large enough to survive in individual pots or the open ground. They usually take several years to reach flowering size, although some bulblets of *Galanthus* have been known to flower in their first year.

Offset

In botany and horticulture, an offset is a small, virtually complete daughter plant that has been naturally and asexually produced on the mother plant. They are clones, meaning that they are genetically identical to the mother plant. In the plant nursery business and gardens, they are detached and grown on in order to produce new plants. This is a cheap and simple process for those plants that readily produce offsets as it does not usually require specialist materials and equipment.

Offsets form when meristem regions of plants, such as axillary buds or homologous structures differentiate into a new plant with the ability to become self-sustaining. This is particularly common in species that develop underground storage organs, such as bulbs, corms and tubers. Tulips and lilies are examples of plants that display offset characteristics by forming cormlets around the original mother corm.

When propagating plants to increase a stock of a cultivar, thus seeking identical copies of parent plant, various cloning techniques (asexual reproduction) are used. Offsets are a natural means by which plants may be cloned.

In contrast, when propagating plants to create new cultivars, sexual reproduction through pollination is used to create seeds. The recombination of genes gives rise to offspring plant with similar but distinct offspring genome.

Apomixis

In botany, apomixis was defined by Hans Winkler as replacement of the normal sexual reproduction by asexual reproduction, without fertilization. Its etymology is Greek for "away from" + "mixing". This definition notably does not mention meiosis. Thus "normal asexual reproduction" of plants, such as propagation from cuttings or leaves, has never been considered to be apomixis, but replacement of the seed by a plantlet or replacement of the flower by bulbils are types of apomixis. Apomictically produced offspring are genetically identical to the parent plant.

Some authors included all forms of asexual reproduction within apomixis, but that generalization of the term has since died out.

In flowering plants, the term "apomixis" is commonly used in a restricted sense to mean agamospermy, i.e. clonal reproduction through seeds. Although agamospermy could theoretically occur in gymnosperms, it appears to be absent in that group.

Apogamy is a related term that has had various meanings over time. In plants with independent gametophytes (notably ferns), the term is still used interchangeably with "apomixis", and both refer to the formation of sporophytes by parthenogenesis of gametophyte cells.

Male apomixis (paternal apomixis) involves replacement of the genetic material of the egg cell by that from the pollen.

Vegetative apomixis in *Poa bulbosa*; bulbils form instead of flowers.

Apomixis and Evolution

Because apomictic plants are genetically identical from one generation to the next, each lineage has some of the characters of a true species, maintaining distinctions from other apomictic lineages within the same genus, while having much smaller differences than is normal between species of most genera. They are therefore often called microspecies.

In some genera, it is possible to identify and name hundreds or even thousands of microspecies, which may be grouped together as species aggregates, typically listed in floras with the convention "*Genus species* agg." (such as the bramble, *Rubus fruticosus* agg.). In some plant families, genera with apomixis are quite common, for example in Asteraceae, Poaceae, and Rosaceae. Examples of apomixis can be found in the genera *Crataegus* (hawthorns), *Amelanchier* (shadbush), *Sorbus* (rowans and whitebeams), *Rubus* (brambles or blackberries), *Poa* (meadow grasses), *Nardus stricta* (Matgrass), *Hieracium* (hawkweeds) and *Taraxacum* (dandelions). Apomixis is reported to occur in about 10% of globally extant ferns. Among polystichoid ferns, apomixis evolved several times independently in three different clades.

Although the evolutionary advantages of sexual reproduction are lost, apomixis can pass along traits fortuitous for evolutionary fitness. As Jens Clausen put it

The apomicts actually have discovered the effectiveness of mass production long before Mr Henry Ford applied it to the production of the automobile. ... Facultative apomixis ... does not prevent variation; rather, it multiplies certain varietal products.

Facultative apomixis means that apomixis does not always occur, i.e. sexual reproduction can also happen. It appears likely that all apomixis in plants is facultative; in other words, that "obligate apomixis" is an artifact of insufficient observation (missing uncommon sexual reproduction).

Apogamy and Apospory in Non-Flowering Plants

The gametophytes of bryophytes, and less commonly ferns and lycopods can develop a group of cells that grow to look like a sporophyte of the species but with the ploidy level of the gametophyte, a phenomenon known as apogamy. The sporophytes of plants of these groups may also have the ability to form a plant that looks like a gametophyte but with the ploidy level of the sporophyte, a phenomenon known as apospory.

Apomixis in Flowering Plants (Angiosperms)

Agamospermy, asexual reproduction through seeds, occurs in flowering plants through many different mechanisms and a simple hierarchical classification of the different types is not possible. Consequently, there are almost as many different usages of terminology for apomixis in angiosperms as there are authors on the subject. For English speakers, Maheshwari 1950 is very influential. German speakers might prefer to consult Rutishauser 1967. Some older text books on the basis of misinformation (that the egg cell in a meiotically unreduced gametophyte can never be fertilized) attempted to reform the terminology to match parthenogenesis as it is used in zoology, and this continues to cause much confusion.

Agamospermy occurs mainly in two forms: In *gametophytic apomixis*, the embryo arises from an unfertilized egg cell (i.e. by parthenogenesis) in a gametophyte that was produced from a cell that did not complete meiosis. In *adventitious embryony* (sporophytic apomixis), an embryo is formed directly (not from a gametophyte) from nucellus or integument tissue.

Types of Apomixis in Flowering Plants

Caribbean Agave producing plantlets on the old flower stem.

Maheshwari used the following simple classification of types of apomixis in flowering plants:

- Nonrecurrent apomixis: In this type "the megaspore mother cell undergoes the usual meiotic divisions and a haploid embryo sac [megagametophyte] is formed. The new embryo may then arise either from the egg (haploid parthenogenesis) or from some other cell of the gametophyte (haploid apogamy)." The haploid plants have half as many chromosomes as the mother plant, and "the process is not repeated from one generation to another" (which is why it is called nonrecurrent).
- Recurrent apomixis, is now more often called gametophytic apomixis: In this type, the megagametophyte has the same number of chromosomes as the mother plant because meiosis was not completed. It generally arises either from an archesporial cell or from some other part of the nucellus.
- Adventive embryony, also called sporophytic apomixis, sporophytic budding, or nucellar embryony: Here there may be a megagametophyte in the ovule, but the embryos do not arise from the cells of the gametophyte; they arise from cells of nucellus or the integument. Adventive embryony is important in several species of *Citrus*, in *Garcinia*, *Euphorbia dulcis*, *Mangifera indica* etc.
- Vegetative apomixis: In this type "the flowers are replaced by bulbils or other vegetative propagules which frequently germinate while **still on** the plant". Vegetative apomixis is important in *Allium*, *Fragaria*, *Agave*, and some grasses, among others.

Types of Gametophytic Apomixis

Gametophytic apomixis in flowering plants develops in several different ways. A megagametophyte develops with an egg cell within it that develops into an embryo through parthenogenesis. The central cell of the megagametophyte may require fertilization to form the endosperm, pseudogamous gametophytic apomixis, or in autonomous gametophytic apomixis fertilization is not required.

- In diplospory (also called generative apospory), the megagametophyte arises from a cell of the archesporium.
- In apospory (also called somatic apospory), the megagametophyte arises from some other nucellus cell.

Considerable confusion has resulted because diplospory is often defined to involve the megaspore mother cell only, but a number of plant families have a multicellular archesporium and the megagametophyte could originate from another archesporium cell.

Diplospory is further subdivided according to how the megagametophyte forms:

- *Allium odorum–A. nutans* type. The chromosomes double (endomitosis) and then meiosis proceeds in an unusual way, with the chromosome copies pairing up (rather than the original maternal and paternal copies pairing up).
- *Taraxacum* type: Meiosis I fails to complete, meiosis II creates two cells, one of which degenerates; three mitotic divisions form the megagametophyte.
- *Ixeris* type: Meiosis I fails to complete; three rounds of nuclear division occur without cell-wall formation; wall formation then occurs.
- *Blumea–Elymus* types: A mitotic division is followed by degeneration of one cell; three mitotic divisions form the megagametophyte.
- *Antennaria–Hieracium* types: three mitotic divisions form the megagametophyte.
- *Eragrostis–Panicum* types: Two mitotic division give a 4-nucleate megagametophyte, with cell walls to form either three or four cells.

Incidence of Apomixis in Flowering Plants

Apomixis occurs in at least 33 families of flowering plants, and has evolved multiple times from sexual relatives. Apomictic species or individual plants often have a hybrid origin, and are usually polyploid.

In plants with both apomictic and meiotic embryology, the proportion of the different types can differ at different times of year, and photoperiod can also change the proportion. It appears unlikely that there are any truly completely apomictic plants, as low rates of sexual reproduction have been found in several species that were previously thought to be entirely apomictic.

The genetic control of apomixis can involve a single genetic change that affects all the major developmental components, formation of the megagametophyte, parthenogenesis of the egg cell, and endosperm development.However, the timing of the various developmental processes is critical to successful development of an apomictic seed, and the timing can be affected by multiple genetic factors.

Some Terms related to Apomixis

- Apomeiosis: "Without meiosis"; usually meaning the production of a meiotically unreduced gametophyte.
- Parthenogenesis: Development of an embryo directly from an egg cell without fertilization is called parthenogenesis. It is of two types:
  - Haploid parthenogenesis: Parthenogenesis of a normal haploid egg (a meiotically reduced egg) into an embryo is termed haploid parthenogenesis. If the mother plant was diploid, then the haploid embryo that results is monoploid, and the plant that grows from the embryo is sterile. If they are not sterile, they are sometimes useful to plant breeders (especially in potato breeding). This type of apomixis has been recorded in *Solanum nigrum*, *Lilium* spp., *Orchis maculata*, *Nicotiana tabacum*, etc.
  - Diploid parthenogenesis: When the megagametophyte develops without completing meiosis, so that the megagametophyte and all cells within it are meiotically unreduced (a.k.a. diploid, but diploid is an ambiguous term), this is called diploid parthenogenesis, and the plant that develops from the embryo will have the same number of chromosomes as the mother plant. Diploid parthenogenesis is a component process of gametophytic apomixis.
- Androgenesis and androclinesis are synonyms. These terms are used for two different processes that both have the effect of producing an embryo that has "male inheritance".

The first process is a natural one. It may also be referred to as male apomixis or paternal apomixis. It involves fusion of the male and female gametes and replacement of the female nucleus by the male nucleus. This has been noted as a rare phenomenon in many plants (e.g. *Nicotiana* and *Crepis*), and occurs as the regular reproductive method in the Saharan Cypress, *Cupressus dupreziana*.

The second process that is referred to as androgenesis or androclinesis involves (artificial) culture of haploid plants from anther tissue or microspores.

- Apogamy: Although this term was (before 1908) used for other types of apomixis, and then discarded as too confusing, it is still sometimes used when an embryo develops from a cell of the megagametophyte other than the egg cell. In flowering plants, the cells involved in apogamy would be synergids or antipodal cells.
- Addition hybrids, called B_{III} hybrids by Rutishauser: An embryo is formed after a meiotically unreduced egg cell is fertilized. The ploidy level of the embryo is therefore higher than that of the mother plant. This process occurs in some plants that are otherwise apomictic, and may play a significant role in producing tetraploid plants from triploid apomictic mother plants (if they receive pollen from diploids). Because fertilization is involved, this process does not fit the definition of apomixis.
- Pseudogamy refers to any reproductive process that requires pollination but does not involve male inheritance. It is sometimes used in a restrictive sense to refer to types of apomixis in which the endosperm is fertilized but the embryo is not. A better term for the restrictive sense is centrogamy.

- Agamospecies, the concept introduced by Göte Turesson: "an apomict population the constituents of which, for morphological, cytological or other reasons, are to be considered as having a common origin," i.e., basically synonymous with "microspecies.

Heated Propagator

A heated propagator is a horticultural device to maintain a warm and damp environment for seeds and cuttings to grow in.

This can be in the form of a clear enclosed bin sitting over a hotpad, or even a portable heater pointed at the bin. The key is to keep the moisture in the clear bin, while keeping lighting over the top of it, usually.

Seed Propagation Mat

An electric seed-propagation mat is a heated rubber mat covered by a metal cage which is used in gardening. The mats are made so that planters containing seedlings can be placed on top of the metal cage without the risk of starting a fire. In extreme cold, gardeners place a loose plastic cover over the planters/mats which creates a sort of miniature greenhouse. The constant and predictable heat allows people to garden in the winter months when the weather is generally too cold for seedlings to survive naturally. When combined with a lighting system, many plants can be grown indoors using these mats.

REFERENCES

1. Nowicki, Marcin; et al. (26 October 2013), More than meets the eye: A multi-year expressivity analyses of tomato sterility in ps and ps-2 lines (PDF), Australian Journal of Crop Science, 7(13):2154-2161; Southern Cross Publishing, retrieved 2013-10-29
2. CHAHAL, G.S & GOSAL, S.S., 2002. Principles and procedures of Plant Breeding, Alpha Science International, United Kingdom.
3. Falconer, D.S., 1989. Introduction to Quantitative Genetics. 3rd Ed. Longman. Burnt Mill.
4. FRISCH, M. & MELCHINGER, A.E., 2005. Selection Theory for Marker-assisted Backcrossing. Genetics: Published Articles Ahead of Print, published on March 31, 2005 as 10.1534/genetics.104.035451
5. GOUSSARD, P.G. (2004) Improving grape vine material through selection. Department of Viticulture. Stellenbosch University. Lecture notes.
6. HOLSINGER, K.E., 2000. Reproductive systems and evolution in vascular plants. PNAS. June 20, vol. 97, no. 13: 7037-7042.\
7. KENNEDY, B.K., 2004. Genetic barrier to self-pollination identified. EurekAlert! 19 May.
8. KOHLI, M.M. & FRANCIS, M., 2000. Application of biotechnologies to wheat

breeding. Proceedings of a conference at La Estanzuela, Uruguay, November 19–20, 1998. Montevideo, Uruguay: CIMMYT.

9. KWV, South Africa. (2005). Setting new global standards for vine plant improvement. Vititec ©
10. MARAIS, G.F. 2005. Seleksiemetodes vir kruisbestuiwende gewasse. Department of Genetics. University of Stellenbosch. Lecture notes.
11. TAKEBAYASHI, N. & MORELL, P.L., 2001. Is self-fertilization an evolutionary dead end? Revisiting an old hypothesis with genetic theories and a macroevolutionary approach. American Journal of Botany. 88:1143-1150.
12. VIVIER, M.A. (2004). Improving grape cultivars. Department of Viticulture. University of Stellenbosch. Lecture notes.
13. Hybrids of different species being sterile, hybrids of same species are not
14. GMO plant made to produce infertile seeds
15. Hottes, A.C. (1925). Practical plant propagation: an exposition of the art and science of increasing plants as practiced by the nurseryman, florist and gardener. translated by. New York: A. T. De La Mare Company, Inc.
16. Principles of Agricultural Botany, p 101, Alexander Nelson, Read Books, 2007, ISBN 1-4067-4662-2
17. Garner, R.J. (1988) The Grafters Handbook P. 131 ISBN 0-304-32172-9
18. http://ccpp.ucr.edu/budwood/budwood.php
19. http://www.freshfromflorida.com/Divisions-Offices/Plant-Industry/Agriculture-Industry/Citrus-Health-Response-Program/Citrus-Budwood-Program
20. http://kcc-weslaco.tamu.edu/budwood.html
21. C.Michael Hogan (2008) Quercus kelloggii, Globaltwitcher, ed. nicklas Stromberg
22. Lang, A., Chailakhyan, M.K. and Frolova, I.A. 1977. Promotion and inhibition of flower formation in a dayneutral plant in grafts with a short-day plant and a long-day plant. Proc Natl Acad Sci U S A. 74(6): 2412-2416.
23. Le Page, Michael (2016-03-17). "Farmers may have been accidentally making GMOs for millennia". The New Scientist. Retrieved 2016-07-11.
24. Nienstaedt, H.; Teich, A. 1972. The genetics of white spruce. USDA, For. Serv., Res. Pap. WO-15. 24 p.
25. Nienstaedt, H. 1966. Dormancy and dormancy release in white spruce. For. Sci. 12:374–384.
26. Nienstaedt, H.; Cech, F.C.; Mergen, F.; Wand, C.; Zak, B. 1958. Vegetative propagation in forest genetics research and practice. J. For. 56:826–839.
27. Greenwood, M.S.; Adams, G.W.; Gillespie, M. 1988. Shortening the breeding cycle of some northeastern conifers. p. 43–52 in Morgenstern, E.K.; Boyle, T.J.B. (Eds.). Tree Improvement – Progressing Together Sympos., Truro NS, Aug. 1987. Proc. Part 2, 21st Meet. Can. Tree Improv. Assoc.
28. Core, J. (2005). "Grafting watermelon onto squash or gourd rootstock makes

firmer, healthier fruit.". Agricultural Research: 53.

29. Onoda, A.; Kobayashi, Ken; Suzuki, Masato (1992), The Study of the Grafting Robot, 319, International Symposium on Transplant Production Systems: International Society for Horticultural Science: Acta Horticulturae, pp. 535–540
30. Kobayashi, Ken; Suzuki, Masato; Sasaya, Sadao (1999), "Grafting Robot", Journal of Robotics and Mechatronics, 11 (3): 213–219
31. "Grafting" (PDF).
32. "Grafting Manual".
33. Cooper and Chapot (1977) Fruit production with special emphasis on fruit for processing. In: Citrus Science and Technology (eds. S. Nagi, P.E. Shur, and M.K. Valdhuis) Vol. 2, p.11
34. Garner, R.J. (1988) The Grafters Handbook P.46 ISBN 0-304-32172-9
35. Hickey, M.; King, C. (2001). The Cambridge Illustrated Glossary of Botanical Terms. Cambridge University Press.
36. "Stolon". Dictionary.com. Retrieved 2007-05-07.
37. Goebel, K.E.v. (1969) [1905]. Organography of plants, especially of the Archegoniatae and Spermaphyta. New York: Hofner publishing company.
38. Gleason, Henry A. (1963). The new Britton and Brown illustrated flora of the Northeastern United States and adjacent Canada, Volume 1. New York: Hafner Press. p. ixxiv. ISBN 0-02-845240-2.
39. Woolley, D. J.; P. F. Wareing (March 1972). "The role of roots, cytokinins and apical dominance in the control of lateral shoot form in Solanum andigena". Planta. 105 (1): 33–42. doi:10.1007/BF00385161. PMID 24477700.
40. Wijesinghe, Dushyantha K.; Dennis F. Whigham (June 2001). "Nutrient foraging in woodland herbs: a comparison of three species of Uvularia (Liliaceae) with contrasting belowground morphologies". American Journal of Botany. 88 (6): 1071–1079. doi:10.2307/2657090. PMID 11410472.
41. Méthy, M.; P. Alpert; J. Roy (September 1990). "Effects of light quality and quantity on growth of the clonal plant Eichhornia crassipes". Oecologia. 84 (2): 265–271. doi:10.1007/BF00318283. Retrieved 2007-05-10.
42. Henderson, Norton C. "Iris". Flora of North America. eFloras.org. Retrieved 2005-05-10.
43. Tucker, Gordon C.; Brian G. Marcks; J. Richard Carter. "Cyperus serotinus". Flora of North America. eFloras.org. Retrieved 2005-05-10.
44. Aflatuni, Abbas; J. Uusitalo; S. Ek; A. Hohtola (January–February 2005). "Variation in the Amount of Yield and in the Extract Composition Between Conventionally Produced and Micropropagated Peppermint and Spearmint". Journal of Essential Oil Research. 17 (1): 66–70. doi:10.1080/10412905.2005. 9698833. ISSN 1041-2905. Retrieved 2005-05-10.
45. Jones, Almut G. (January 1978). "Observations on Reproduction and Phenology in Some Perennial Asters". American Midland Naturalist. 99 (1): 184–97.

doi:10.2307/2424942. JSTOR 2424942.

46. Visser, Richard G. F.; Dick Vreugdenhil; Theo Hendriks; Evert Jacobsen (February 1994). "Gene expression and carbohydrate content during stolon to tuber transition in potatoes (Solanum tuberosum)". Physiologia Plantarum. 90 (2): 285–92. doi:10.1111/j.1399-3054.1994.tb00389.x.
47. Monaco Educational Service. "Introduction to stems". Botany. Archived from the original on 2005-04-13. Retrieved 2005-05-10.
48. Hartmann, Hudson Thomas; Dale E Kester (1983). Plant propagation : principles and practices. Englewood Cliffs: Prentice-Hall. p. 508. ISBN 0-13-681007-1.
49. "Hydrilla in the Catawba River Basin" (PDF). NCSU Aquatic Weed Management Program. Retrieved 2005-05-10.
50. weakley, Alan S. "Flora of the Carolinas, Virginia, and Georgia, and Surrounding Areas, Part 6" (PDF). p. 808. Retrieved 2005-05-10.
51. Harris, Peter. "Field and Hedge bindweeds Convolvulus arvensis L. and Calystegia sepium (L.) R. Br.". Classical Biological Control of Weeds. Agriculture and Agri-Food Canada, Lethbridge Research Centre. Retrieved 2005-05-10.
52. Pierdominici, Maria Grazia; Ming Dong (January 1995). "Morphology and growth of stolons and rhizomes in three clonal grasses, as affected by different light supply". Plant Ecology. 116 (1): 25–32. doi:10.1007/BF00045274. Retrieved 2005-05-10.
53. Levinton, Jeffrey S. "Marine Biology." 3rd Edition. Oxford Press. 2008.
54. "Polychaetous worm with engine and wagons" by Fredrik Pleijel|
55. Mitchell et al. (2015), doi:10.1038/nature14646
56. Peterson et al. (2003), Integr Comp Biol, 43:127-36
57. Winkler, H. (1908). "Über Parthenogenesis und Apogamie im Pflanzenreich". Progressus Rei Botanicae. 2 (3): 293–454.
58. Ross A. Bicknell; Anna M. Koltunow (2004). "Understanding Apomixis: Recent Advances and Remaining Conundrums". The Plant Cell. 16 (suppl 1): S228–S245. doi:10.1105/tpc.017921.
59. Hong-Mei Liu, Robert J. Dyer, Zhi-You Guo, Zhen Meng,Jian-Hui Li, and Harald Schneider. (2012) The Evolutionary Dynamics of Apomixis in Ferns: A Case Study from Polystichoid Ferns. Journal of Botany Volume 2012 Article ID 510478, 11 pages http://dx.doi.org/10.1155/2012/510478
60. Clausen, J. (1954). "Partial apomixis as an equilibrium system". Caryologia. 1954, Supplement: 469–479.
61. Savidan, Y.H. (2000). "Apomixis: genetics and breeding". Plant Breeding Reviews. 18: 13–86. doi:10.1002/9780470650158.ch2.
62. Steil, W.N. (1939). "Apogamy, apospory, and parthenogenesis in the Pteridophytes". The Botanical Review. 5 (8): 433–453. doi:10.1007/bf02878704.
63. Niklas, K.J. (1997). The evolutionary biology of plants. Chicago: The University of Chicago press.

64. Maheshwari, P. 1950. An introduction to the embryology of the angiosperms. McGraw-Hill, New York.

65. Rutishauser, A. 1969. Embryologie und Fortpflanzungsbiologie der Angiospermen: eine Einführung. Springer-Verlag, Wien.

66. Fitting, H., et al. 1930. Textbook of botany (Strasburger's textbook of botany, rewritten). Macmillan, London.

67. Nogler, G.A. 1984. Gametophytic apomixis. In Embryology of angiosperms. Edited by B.M. Johri. Springer, Berlin, Germany. pp. 475–518.

68. Carman, J.G. (1997). "Asynchronous expression of duplicate genes in angiosperms may cause apomixis, bispory, tetraspory, and polyembryony". Biological Journal of the Linnean Society. 61 (1): 51–94. doi:10.1111/j.1095-8312.1997.tb01778.x.

69. Nygren, A. (1967). "Apomixis in the angiosperms". In W. Ruhland. Handbuch der Pflanzenphysiologie. 18. Berlin: Springer-Verlag. pp. 551–596.

70. Koltunow, A.M.; Johnson, S.D.; Bicknell, R.A. (2000). "Apomixis is not developmentally conserved in related, genetically characterized Hieracium plants of varying ploidy". Sexual Plant Reproduction. 12 (5): 253–266. doi:10.1007/s004970050193.

71. Christian Pichot; Benjamin Liens; Juana L. Rivera Nava; Julien B. Bachelier; Mohamed El Maâtaoui (2008). %7C doi: 10.1534/genetics.107.080572 Genetics January 2008 vol. 178 no. 1 379-383 "Cypress Surrogate Mother Produces Haploid Progeny From Alien Pollen" Check |url= value (help). Genetics. 178 (1): 379–383. doi:10.1534/genetics.107.080572.

72. Christian Pichot; Bruno Fady; Isabelle Hochu (2000). "Lack of mother tree alleles in zymograms of Cupressus dupreziana A. Camus embryos" (PDF). Annals of Forest Science. 57: 17–22. doi:10.1051/forest:2000108.

73. Pichot, C.; El Maataoui, M.; Raddi, S.; Raddi, P. (2001). "Conservation: Surrogate mother for endangered Cupressus". Nature. 412 (6842): 39. doi:10.1038/35083687.

74. Solntzeva, M.P. (2003). "About some terms of apomixis: pseudogamy and androgenesis". Biologia. 58 (1): 1–7.

75. Defining species: a sourcebook from antiquity to today, by John S. Wilkins, ISBN 1433102161, 2009, pp. 122, 194

76. Gvaladze G.E. (1976). Forms of Apomixis in the genus Allium L. In: S.S. Khokhlov (Ed.): Apomixis and Breeding, Amarind Pub., New Delhi-Bombay-Calcutta-New York pp. 160–165

77. Bhojwani S.S.& Bhatnagar S.P. (1988). The Embryology of angiosperms. Vikas Publishing house Pvt.Ltd. New Delhi.

78. Heslop-Harrison, J. (1972) "Sexuality in Angiosperms,"pp. 133–289, In Steward,F.C. (ed.) Plant Physiology, Vol. 6C, Academic Press New York.

CHAPTER 5

Smart Breeding

Chapter Contents

- **Introduction**
- **Genomics of Domestication**
- **Genomics as a Tool**
- **Genetics and Genomics of Domestication**

Smart Breeding is a method based on the results of genomic research. Smart Breeding is precision breeding: When selecting the plants, the breeders do not rely on outward features (phenotype). Instead, the genetics are analyzed in detail (genomic research) in order to either select suitable partners for crossing or offspring featuring the desirable traits. For this, plant breeders use molecular diagnostic tools - the so-called molecular markers. In contrast to Genetic Engineering, Smart Breeding does not introduce any new genes into the genome. The resulting organisms are not transgenic.

Smart Breeding enables the detection of desired performance features irrespective of environmental influences. The analysis of the trait in question is a quick and easily automated process allowing a high throughput. Moreover, the existence of a trait can be analyzed in earlier generations and in earlier development states (germ bud). Traditional methods can only decide upon resistance by carrying out infection trials in breeding nurseries or laboratories and by analyzing yield and quality of the harvested crop. Genomic research can supplant such time- and cost-intensive laboratory or field tests and can speed up the selection process. All in all, breeding results have become much more predictable.

INTRODUCTION

SMART breeding (Selection with Markers and Advanced Reproductive Technologies) or Precision breeding refers to a genetic engineering technique of reproducing a species members together to retain desirable traits and so produce a stronger hybrid. The technique was successfully used by Nachum Kedar, an Israeli scientist, who applied the technique using beefsteak tomatoes to produce a fruit that would ripen on the vine and remain firm in transit.

Smart breeding has been advanced as an alternative to transgenics as a way to produce plants that are resistant to various environmental problems.

GENOMICS OF DOMESTICATION

Domesticated species and the human populations that domesticate them are typified by a mutualistic relationship of interdependence, in which humans have over thousands of years modified the genomics of domesticated species. Genomics is the study of the structure, content, and evolution of genomes, or the entire genetic information of organisms. Domestication is the process by which humans alter the morphology and genes of targeted organisms by selecting for desirable traits.

Since domestication involves selection of traits over time, which leads to genetic changes, the science of genomics can identify which genes across an entire genome are altered during this intense artificial selection period. Understanding the genomics of domestication can also offer insight into the genetic effects of both the artificial, human driven selection of domestication, as well as natural selection. This makes the genomics of domestication a unique tool for examining the genetics of evolution in organisms that are relatively easy to study as their history may be more thoroughly preserved due to their usefulness to humans.

GENOMICS AS A TOOL

Historically genomic studies have been focused on select organisms for which there is funding to study. Initially, when sequencing costs were prohibitive, this was limited to organisms with small genomes, such as viruses and bacteria, and then in eukaryotes, model organisms of importance to the scientific community for research. These included the *Mus musculus* (the house mouse), *Drosophila melanogaster* (fruit fly) and *Arabidopsis thaliana* (*Arabidopsis*) genomes. One of the most prominent publicly funded genome projects was the Human Genome Project which helped to refine existing sequencing techniques as well as develop additional ones. Following these model organisms, agriculturally important species were next emphasized. As of 2009, there are more than 50 plant species whose genomes are being sequenced. However, the most important agricultural crops, including those in the grass and legume families such as rice, wheat and maize, have received the most attention and funding. As of 2005, a full sequence of the rice genome has been published. These domesticated species and in some cases, their wild ancestors, have received focus due to their agricultural and economic importance and the benefits that having a sequenced genome for these species confers such as the ability to easily identify targets for selective breeding programs to increase yield, facilitate drought tolerance, or select a variety of desirable traits.

Diversity of ancestral rice.

GENETICS AND GENOMICS OF DOMESTICATION

During domestication, crop species undergo intense selective pressures that alter their genomes. The process of selection during domestication has largely focused on core traits that have come to define domesticated species. In seed or grain crops, these hallmark traits include increases in seed size, a reduction in natural seed dispersal, reduced lateral branching, and an annual life cycle. The genes that code for these traits have been elucidated in some species, such as the maize tb1 gene, which controls for lateral branching, using classical genetic techniques as well as genomics. However, traditional Mendelian genetics which examines inheritance patterns on an individual trait basis is limited to traits or phenotypes that cleanly segregate into distinct classes. Genomics is able to overcome this limitation through the comparison of the genomes of individuals exhibiting a trait or phenotype of interest to a reference genome which enables the identification differences between the two genomes such as single-nucleotide polymorphisms (SNP), the movement of transposable elements (or retrotransposons) or deletions, among other genetic changes.

Coding DNA

Genomics offers insight into coding DNA as well as noncoding DNA. By comparing the sequence of a previously isolated section of chromosome 8 in rice between fragrant and non-fragrant varietals researchers were able to determine their genetic difference. The aromatic and fragrant rices, including basmati and jasmine are derived from an ancestral rice domesticate that suffered a deletion in exon 7 and as a result sequence coding for betaine aldehyde dehydrogenase (BADH2) was altered.

Noncoding DNA

However, looking solely at genes, or coding DNA, can be ineffective when examining certain traits or studying the evolution of a species during the domestication process. Genes that are vital for cellular process are often highly conserved and mutations at these locations can prove fatal. Areas of the genome that are noncoding can be prone to much higher mutation rates. Because of this, these noncoding genes provide vital information when studying the divergence of wild and domestic species. Since core genes are conserved between and among species, examining DNA sequences for these genes in multiple individuals of a species may be unable to provide much information on the diversity present in a population or species that is young. The estimated age of domesticated animal and plant species tends to be less than 10,000 years, which on an evolutionary timescale, is relatively short. Because of this, highly variable noncoding DNA, such as microsatellites, that mutate frequently, provide genetic markers with sufficient intraspecific variation to document domestication. Studying the noncoding DNA of domesticated species is made possible by genomics, which provides the genetic sequence of the entire genome, not simply coding DNA from genes of interest. In the case of coconuts, recent genomic research using 10 microsatellite loci was able to determine that there have been 2 cases of coconut domestication based on sufficient variation between individuals found in the Indian Ocean and those found in the Pacific Ocean.

Advantages of Genomics over Traditional Genetics

Genomics offers various benefits that the study of single genes, or genetics, does not. Having a fully sequenced genome for an organism, such as potato, allows researchers to compare DNA across multiple species and examine conserved sequences. In 2011, researchers at the Potato Genome Sequencing Consortium compared a de novo potato genome to 12 other species including arabidopsis, grape, rice, sorghum, maize, poplar, and others that allowed them to isolate potato specific genes, including those that confer resistance to potato blight caused by *Phytophthora infestans*. The ability to predict genes of interest for crop breeding is a major advantage to the further domestication of crop species that is facilitated by genomics and the identification of genes and extragenic sequences that control for these desirable traits. Modern plant breeders can use this information to manipulate the genetics of crop species to develop new domesticated varieties with desired modern traits such as increased yield and the ability to respond better to nitrogen fertilizers. Comparative genomics also allows researchers to make inferences about the evolution of life through comparing genomic sequences and examining patterns of divergence and conservation.

Evolution

In his most famous work, *Origin of Species*, Charles Darwin compared natural selection to domestication to help explain the former and he went on to write an entire book on the topic entitled *The Variation of Animals and Plants under Domestication*. Domesticated species serve as ideal model systems for examining key concepts of evolution because their history is relatively short (on the evolutionary scale of billions of years) and well preserved. Additionally, by virtue of their usefulness to humans, many domesticated species are extant and available for study. The genomes of crop species have been sequenced in part to help with their improvement for agronomic reasons, but because genome data are publicly available, in many cases for free, these organisms also serve as systems for examining the effects of evolution and artificial selection on genes. In particular, the genomics of domesticated species allow for the study of strong artificial selection, founder events and bottlenecks, as well as wider evolutionary questions.

The process of domestication, by which only a choice few wild individuals are cultivated and selected against, often results in very strong selective pressures. This is evident in the genomes of these individuals as a lack of genetic diversity. In some cases this lack of diversity is seen as a selective sweep, whereby the variation at a particular locus of the genome is highly reduced while variation outside of this area is maintained or only partially reduced. In other cases, such as the coconut, genomic studies have revealed occurrences of a founder event, whereby the genetic diversity of an entire population is reduced due to a small number of individuals with low diversity being the isolated ancestors of a larger modern population. Bottlenecks, where variation is reduced throughout the entire genome, are also evident in crop species such as pearl millet, cotton, common bean and lima bean. With the identification of bottlenecks in these species, researchers are able to study the effects on an organisms' ability to evolve past a bottleneck, and what effects this may have on the genomes of both individuals and populations as well as their fitness.

Domesticated species and human history

Domesticated species and the human populations that domesticate them are typified by a mutualistic relationship of interdependence. Domesticated crop species tend to become increasingly reliant on human populations for dispersal due to the selection against natural seed dispersal methods and humans have become increasingly dependent on domesticated crop species to sustain growing populations. Because many crop species rely on humans for dispersal, and it is possible to use genomics to track the dispersal of domesticated species, the genomics of domesticated species can be used as a tool to track human movements throughout history.

Bottle Gourd

The bottle gourd (*lagenaria siceraria*) is a domesticated species that originated in Africa and was dispersed throughout Asia by 9000 B.C.E. and reached the Americas by 8000 B.C.E. Morphologically and genetically, the Asian and African bottle gourds are sufficiently different that they can be designated as two separate subspecies. Morphologically the American gourd resembles the African gourds more than the Asian gourds, which was previously used as support for the theory that the American variety is derived from a wild African gourd that floated across the ocean. However, in 2005 researchers with the Smithsonian Institution were able to use a combination of archeological and genomic data to show that the bottle gourds in the Americas are actually more similar to Asian gourds, which suggests that the American gourds may be derived from Asian gourds that were carried across the Bering land bridge by Paleo-Indians.

Coconut

Genomic analysis of cultivated coconut (*Cocos nucifera*) has shed light on the movements of Austronesian peoples. By examining 10 microsatelite loci, researchers found that there are 2 genetically distinct subpopulations of coconut – one originating in the Indian Ocean, the other in the Pacific Ocean. However, there is evidence of admixture, the transfer of genetic material, between the two populations. Given that coconuts are ideally suited for ocean dispersal, it seems possible that individuals from one population could have floated to the other. However, the locations of the admixture events are limited to Madagascar and coastal east Africa and exclude the Seychelles. This pattern coincides with the known trade routes of Austronesian sailors. Additionally, there is a genetically distinct sub-population of coconuts on the Pacific coast of Latin America which has undergone a genetic bottleneck, resulting from a founder effect; however, its ancestral population is the pacific coconut, which suggests that Austronesian peoples may have sailed as far east as the Americas.

REFERENCES

1. Gibson, Greg (2009). A Primer of Genome Science. Sunderland, MA: Sinauer Associates. ISBN 978-0-87893-236-8.
2. Purugganan, Michael; Dorian Fuller (February 2009). "The nature of selection during plant domestication". Nature. 12. 457 (7231): 843–8. doi:10.1038/nature07895. PMID 19212403. Retrieved 28 November 2011.
3. Gepts, Paul (2004). "Crop Domestication as a long-term selection experiment" (PDF). Plant Breeding Reviews. 2. 24. Retrieved 28 November 2011.
4. Shao, G; A. Tang; S. Q. Tang; J. Luo; G. A. Jiao; J. L. Wu; P. S. Hu (April 2011). "A new deletion mutation of fragrant gene and the development of three molecular markers for fragrance in rice". Plant Breeding. 2. 130: 172–176. doi:10.1111/j.1439-0523.2009.01764.x.
5. Zeder, Melinda; Eve Emshwiller; Bruce D. Smith; Daniel G. Bradley (March 2006). "Documenting domestication: the intersection of genetics and archaeology". Trends in Genetics. 22 (3): 139–55. doi:10.1016/j.tig.2006.01.007. PMID 16458995. Retrieved 28 November 2011.
6. Gunn, Bee; Luc Baudouin; Kenneth M. Olsen (2011). "Independent Origins of Cultivated Coconut (Cocos nucifera L.) in the Old World Tropics". PLoS

ONE. 6 (6): e21143. doi:10.1371/journal.pone.0021143. PMC 3120816Freely accessible. PMID 21731660. Retrieved 28 November 2011.

7. The Potato Genome Sequencing Consortium (July 2011). "Genome sequence and analysis of the tuber crop potato". Nature. 475 (7355): 189–95. doi:10.1038/nature10158. PMID 21743474. Retrieved 28 November 2011.
8. Ross-Ibarra, Jeffery; Peter L. Morrell; Brandon S. Gaut (May 2007). "Plant domestication, a unique opportunity to identify the genetic basis of adaptation". PNAS. 104: 8641–8648. doi:10.1073/pnas.0700643104. PMC 1876441Freely accessible. PMID 17494757. Retrieved 28 November 2011.
9. Erickson, David; Bruce D. Smith; Andrew C. Clarke; Daniel H. Sandweiss; Noreen Tuross (2005). "An Asian origin for a 10,000-year-old domesticated plant in the Americas". PNAS. 102 (51): 18315–18320. doi:10.1073/pnas.0509279102. PMC 1311910Freely accessible. PMID 16352716. Retrieved 28 November 2011.

CHAPTER 6

Breeding in Self-Pollination Plants

Chapter Contents

INTRODUCTION

Pollination is the process by which pollen is transferred to the female reproductive organs of a plant, thereby enabling fertilization to take place. Like all living organisms, seed plants have a single major goal: to pass their genetic information on to the next generation. The reproductive unit is the seed, and pollination is an essential step in the production of seeds in all spermatophytes (seed plants).

For the process of pollination to be successful, a pollen grain produced by the anther, the male part of a flower, must be transferred to a stigma, the female part of the flower, of a plant of the same species. The process is rather different in angiosperms (flowering plants) from what it is in gymnosperms (other seed plants). In angiosperms, after the pollen grain has landed on the stigma, it creates a pollen tube which grows down the style until it reaches the ovary. Sperm cells from the pollen grain then move along the pollen tube, enter the egg cell through the micropyle and fertilise it, resulting in the production of a seed.

A successful angiosperm pollen grain (gametophyte) containing the male gametes is transported to the stigma, where it germinates and its pollen tube grows down the style to the ovary. Its two gametes travel down the tube to where the gametophyte(s) containing the female gametes are held within the carpel. One nucleus fuses with the polar bodies to produce the endosperm tissues, and the other with the ovule to produce the embryo Hence the term: "double fertilization".

In gymnosperms, the ovule is not contained in a carpel, but exposed on the surface of a dedicated support organ, such as the scale of a cone, so that the penetration of carpel tissue is unnecessary. Details of the process vary according to the division of gymnosperms in question. Two main modes of fertilization are found in gymnosperms. Cycads and *Ginkgo* have motile sperm that swim directly to the egg inside the ovule, whereas conifers and gnetophytes have sperm that are unable to swim but are conveyed to the egg along a pollen tube.

The study of pollination brings together many disciplines, such as botany, horticulture, entomology, and ecology. The pollination process as an interaction between flower and pollen vector was first addressed in the 18th century by Christian Konrad Sprengel. It is important in horticulture and agriculture, because fruiting is dependent on fertilization: the result of pollination. The study of pollination by insects is known as *anthecology*.

Carpenter bee with pollen collected from Night-blooming cereus

Tip of a tulip stamen covered with pollen grains.

POLLINATION PROCESS

Pollen germination has three stages; hydration, activation and pollen tube emergence. The pollen grain is severely dehydrated so that its mass is reduced enabling it to be more easily transported from flower to flower. Germination only takes place after rehydration, ensuring that premature germination does not take place in the anther. Hydration allows the plasma membrane of the pollen grain to reform into its normal bilayer organization providing an effective osmotic membrane. Activation involves the development of actin filaments throughout the cytoplasm of the cell, which eventually become concentrated at the point from which the pollen tube will emerge. Hydration and activation continue as the pollen tube begins to grow.

In conifers, the reproductive structures are borne on cones. The cones are either pollen cones (male) or ovulate cones (female), but some species are monoecious and others dioecious. A pollen cone contains hundreds of microsporangia carried on (or borne on) reproductive structures called sporophylls. Spore mother cells in the microsporangia divide by meiosis to form haploid microspores that develop further by two mitotic divisions into immature male gametophytes (pollen grains). The four resulting cells consist of a large tube cell that forms the pollen tube, a generative cell that will produce two sperm by mitosis, and two prothallial cells that degenerate. These cells comprise a very reduced microgametophyte, that is contained within the resistant wall of the pollen grain.

The pollen grains are dispersed by the wind to the female, ovulate cone that is made up of many overlapping scales (sporophylls, and thus megasporophylls), each protecting two ovules, each of which consists of a megasporangium (the nucellus) wrapped in two layers of tissue, the integument and the cupule, that were derived from highly modified branches of ancestral gymnosperms. When a pollen grain lands close enough to the tip of an ovule, it is drawn in through the micropyle (a pore in the integuments covering the tip of the ovule) often by means of a drop of liquid known as a pollination drop. The pollen enters a pollen chamber close to the nucellus, and there it may wait for a year before it germinates and forms a pollen tube that grows through the wall of the megasporangium (=nucellus) where fertilisation takes place. During this time, the megaspore mother cell divides by meiosis to form four haploid cells, three of which degenerate. The surviving one develops as a megaspore and divides repeatedly to form an immature female gametophyte (egg sac). Two or three archegonia containing an egg then develop inside the gametophyte. Meanwhile, in the spring of the second year two sperm cells are produced by mitosis of the body cell of the male gametophyte. The pollen tube elongates and pierces and grows through the megasporangium wall and delivers the sperm cells to the female gametophyte inside. Fertilisation takes place when the nucleus of one of the sperm cells enters the egg cell in the megagametophyte's archegonium.

In flowering plants, the anthers of the flower produce microspores by meiosis. These undergo mitosis to form male gametophytes, each of which contains two haploid cells. Meanwhile, the ovules produce megaspores by meiosis, further division of these form the female gametophytes, which are very strongly reduced, each consisting only of a few cells, one of which is the egg. When a pollen grain adheres to the stigma of a carpel it germinates, developing a pollen tube that grows through the tissues of the style, entering the ovule through the micropyle. When the tube reaches the egg sac, two sperm cells pass through it into the female gametophyte and fertilisation takes place.

TYPES

Abiotic

Abiotic pollination refers to situations where pollination is mediated without the involvement of other organisms. The most common form of abiotic pollination, anemophily, is pollination by wind. Wind pollination is very imprecise, with a minute proportion of pollen grains landing by chance on a suitable receptive stigma, the rest being wasted in the environment. This form of pollination is used by grasses, most conifers, and many deciduous trees. Hydrophily is pollination by water, and occurs in aquatic plants which release their pollen directly into the surrounding water. About 80% of all plant pollination is biotic. In gymnosperms, biotic pollination is generally incidental when it occurs, though some gymnosperms and their pollinators are mutually adapted for pollination. The best-known examples probably are members of the order Cycadales and associated species of beetles. Of the abiotically pollinated species of plant, 98% are anemophilous and 2% hydrophilous, their pollen being transported by water.

It is thought that among angiosperms, entomophily is the primitive state; this is indicated by the vestigial nectaries in the wind-pollinated *Urtica* and other plants, and the presence of fragrances in some of these plants. Of the angiosperms, grasses, sedges, rushes and catkin-bearing plants are in general wind pollinated. Other flowering plants are mostly biotic, the pollen being carried by animal vectors. However a number of plants in multiple families have secondarily adopted wind pollination in contrast to other members of their groups. Some plants are intermediate between the two pollination methods. common heather is regularly pollinated by insects, but produce clouds of pollen and some wind pollination is inevitable, and the hoary plantain is primarily wind pollinated, but is also visited by insects which pollinate it.

Melissodes desponsa covered in pollen.

Biotic

More commonly, the process of pollination requires pollinators: organisms that carry or move the pollen grains from the anther of one flower to the receptive part of the carpel or pistil (stigma) of another. This is biotic pollination. The various flower traits (and combinations thereof) that differentially attract one type of pollinator or another are known as pollination syndromes. At least 100,000 species of animal, and possibly as many as 200,000, act as pollinators of the estimated 250,000 species of flowering plants in the world. The majority of these pollinators are insects, but about 1,500 species of birds and mammals have been reported to visit flowers and may transfer pollen between them. Besides birds and bats which are the most frequent visitors, these include monkeys, lemurs, squirrels, rodents and possums.

Hummingbirds typically feed on red flowers

Entomophily, pollination by insects, often occurs on plants that have developed colored petals and a strong scent to attract insects such as, bees, wasps and occasionally ants (Hymenoptera), beetles (Coleoptera), moths and butterflies (Lepidoptera), and flies (Diptera). The existence of insect pollination dates back to the dinosaur era.

In zoophily, pollination is performed by vertebrates such as birds and bats, particularly, hummingbirds, sunbirds, spiderhunters, honeyeaters, and fruit bats. Plants adapted to using bats or moths as pollinators typically have white petals and a strong scent and flower at night, whereas plants that use birds as pollinators tend to produce copious nectar and have red petals.

Insect pollinators such as honeybees (*Apis mellifera*), bumblebees (*Bombus terrestris*), and butterflies (*Thymelicus flavus*) have been observed to engage in flower constancy, which means they are more likely to transfer pollen to other conspecific plants. This can be beneficial for the pollinators, as flower constancy prevents the loss of pollen during interspecific flights and pollinators from clogging stigmas with pollen of other flower species. It also improves the probability that the pollinator will find productive flowers easily accessible and recognisable by familiar clues.

MECHANISM

A European honey bee collects nectar, while pollen collects on its body.

Africanized honey bees immersed in Yellow Opuntia engelmanniiCactus Flower Pollen

Pollination can be accomplished by cross-pollination or by self-pollination:

- Cross-pollination, also called *allogamy*, occurs when pollen is delivered from the stamen of one flower to the stigma of a flower on another plant of the same species. Plants adapted for cross-pollination have several mechanisms to prevent self-pollination; the reproductive organs may be arranged in such a way that self-fertilization is unlikely, or the stamens and carpels may mature at different times.
- Self-pollination occurs when pollen from one flower pollinates the same flower or other flowers of the same individual. It is thought to have evolved under

conditions when pollinators were not reliable vectors for pollen transport, and is most often seen in short-lived annual species and plants that colonize new locations. Self-pollination may include *autogamy*, where pollen is transferred to the female part of the same flower; or *geitonogamy*, when pollen is transferred to another flower on the same plant. Plants adapted to self-fertilize often have similar stamen and carpel lengths. Plants that can pollinate themselves and produce viable offspring are called self-fertile. Plants that cannot fertilize themselves are called self-sterile, a condition which mandates cross-pollination for the production of offspring.

- *Cleistogamy*: is self-pollination that occurs before the flower opens. The pollen is released from the anther within the flower or the pollen on the anther grows a tube down the style to the ovules. It is a type of sexual breeding, in contrast to asexual systems such as apomixis. Some *cleistogamous* flowers never open, in contrast to *chasmogamous* flowers that open and are then pollinated. Cleistogamous flowers are by necessity found on self-compatible or self-fertile plants. Although certain orchids and grasses are entirely cleistogamous, other plants resort to this strategy under adverse conditions. Often there may be a mixture of both cleistogamous and chasmogamous flowers, sometimes on different parts of the plant and sometimes in mixed inflorescences. The ground bean produces cleistogamous flowers below ground, and mixed cleistogamous and chasmogamous flowers above.

Geranium incanum, like most geraniums and pelargoniums, sheds its anthers, sometimes its stamens as well, as a barrier to self-pollination. This young flower is about to open its anthers, but has not yet fully developed its pistil.

These *Geranium incanum* flowers have opened their anthers, but not yet their stigmas. Note the change of color that signals to pollinators that it is ready for visits.

This *Geranium incanum* flower has shed its stamens, and deployed the tips of its pistil without accepting pollen from its own anthers. (It might of course still receive pollen from younger flowers on the same plant.)

An estimated 48.7% of plant species are either dioecious or self-incompatible obligate out-crossers. It is also estimated that about 42% of flowering plants exhibit a mixed mating system in nature. In the most common kind of mixed mating system, individual plants produce a single type of flower and fruits may contain self-pollinated, out-crossed or a mixture of progeny types.

Pollination also requires consideration of pollenizers. The terms "pollinator" and "pollenizer" are often confused: a *pollinator* is the agent that moves the pollen, whether it be bees, flies, bats, moths, or birds; a *pollenizer* is the plant that serves as the pollen source for other plants. Some plants are *self-compatible* (*self-fertile*) and can pollinate and fertilize themselves. Other plants have chemical or physical barriers to self-pollination.

In agriculture and horticulture pollination management, a good pollenizer is a plant that provides compatible, viable and plentiful pollen and blooms at the same time as the plant that is to be pollinated or has pollen that can be stored and used when needed to pollinate the desired flowers. Hybridization is effective pollination between flowers of different species, or between different breeding lines or populations.

Peaches are considered self-fertile because a commercial crop can be produced without cross-pollination, though cross-pollination usually gives a better crop. Apples are considered self-incompatible, because a commercial crop must be cross-pollinated. Many commercial fruit tree varieties are grafted clones, genetically identical. An orchard block of apples of one variety is genetically a single plant. Many growers now consider this a mistake. One means of correcting this mistake is to graft a limb of an appropriate pollenizer (generally a variety of crabapple) every six trees or so.

The wasp *Mischocyttarus* rotundicollis transporting pollen grains of *Schinus terebinthifolius*

Pollen Vectors

Biotic pollen vectors are animals, usually insects, but also reptiles, birds, mammals, and sundry others, that routinely transport pollen and play a role in pollination. This is usually as a result of their activities when visiting plants for feeding, breeding or shelter. The pollen adheres to the vector's body parts such as face, legs, mouthparts, hair, feathers, and moist spots; depending on the particular vector. Such transport is vital to the pollination of many plant species.

Any kind of animal that often visits or encounters flowers is likely to be a pollen vector to some extent. For example, a crab spider that stops at one flower for a time and then moves on, might carry pollen incidentally, but most pollen vectors of significant interest are those that routinely visit the flowers for some functional activity. They might feed on pollen, or plant organs, or on plant secretions such as nectar, and carry out acts of pollination on the way. Many plants bear flowers that favour certain types of pollinator over all others. This need not always be an effective strategy, because some flowers that are of such a shape that they favor pollinators that pass by their anthers and stigmata on the way to the nectar, may get robbed by ants that are small enough to bypass the normal channels, or by short-tongued bees that bite through the bases of deep corolla tubes to extract nectar at the end opposite to the anthers and stigma. Some flowers have specialized mechanisms to trap pollinators to increase effectiveness. Other flowers will attract pollinators by odor. For example, bee species such as *Euglossa cordata* are attracted to orchids this way, and it has been suggested that the bees will become intoxicated during these visits to the orchid flowers, which last up to 90 minutes. However, in general, plants that rely on pollen vectors tend to be adapted to their particular type of vector, for example day-pollinated species tend to be brightly coloured, but if they are pollinated largely by birds or specialist mammals, they tend to be larger and have larger nectar rewards than species that are strictly insect-pollinated. They also tend to spread their rewards over longer periods, having long flowering seasons; their specialist pollinators would be likely to starve if the pollination season were too short.

As for the types of pollinators, reptile pollinators are known, but they form a minority in most ecological situations. They are most frequent and most ecologically significant in island systems, where insect and sometimes also bird populations may be unstable and less species-rich. Adaptation to a lack of animal food and of predation pressure, might therefore favor reptiles becoming more herbivorous and more inclined to feed on pollen and nectar. Most species of lizards in the families that seem to be significant in pollination seem to carry pollen only incidentally, especially the larger species such as Varanidae and Iguanidae, but especially several species of the Gekkonidae are active pollinators, and so is at least one species of the Lacertidae, *Podarcis lilfordi*, which pollinates various species, but in particular is the major pollinator of *Euphorbia dendroides* on various Mediterranean islands.

Mammals are not generally thought of as pollinators, but some rodents, bats and marsupials are significant pollinators and some even specialize in such activities. In South Africa certain species of *Protea* (in particular *Protea humiflora*, *P. amplexicaulis*, *P. subulifolia*, *P. decurrens* and *P. cordata*) are adapted to pollination by rodents (particularly Cape Spiny Mouse, *Acomys subspinosus*) and elephant shrews (*Elephantulus* species). The flowers are borne near the ground, are yeasty smelling, not colourful, and sunbirds reject the nectar with its high xylose content. The mice apparently can digest the xylose and they eat large quantities of the pollen. In Australia pollination by flying, gliding and earthbound mammals has been demonstrated.

Examples of pollen vectors include many species of wasps, that transport pollen of many plant species, being potential or even efficient pollinators.

EVOLUTION OF PLANT/POLLINATOR INTERACTIONS

The first fossil record for abiotic pollination is from fern-like plants in the late Carboniferous period. Gymnosperms show evidence for biotic pollination as early as the Triassic period. Many fossilized pollen grains show characteristics similar to the biotically dispersed pollen today. Furthermore, the gut contents, wing structures, and mouthpart morphologies of fossilized beetles and flies suggest that they acted as early pollinators. The association between beetles and angiosperms during the early Cretaceous period led to parallel radiations of angiosperms and insects into the late Cretaceous. The evolution of nectaries in late Cretaceous flowers signals the beginning of the mutualism between hymenopterans and angiosperms.

Bees provide a good example of the mutualism that exists between hymenopterans and angiosperms. Flowers provide bees with nectar (an energy source) and pollen (a source of protein). When bees go from flower to flower collecting pollen they are also depositing pollen grains onto the flowers, thus pollinating them. While pollen and nectar, in most cases, are the most notable reward attained from flowers, bees also visit flowers for other resources such as oil, fragrance, resin and even waxes. It has been estimated that bees originated with the origin or diversification of angiosperms. In addition, cases of coevolution between bee species and flowering plants have been illustrated by specialized adaptations. For example, long legs are selected for in *Rediviva neliana*, a bee that collects oil from *Diascia capsularis*, which have long spur lengths that are selected for in order to deposit pollen on the oil-collecting bee, which in turn selects for even longer legs in *R. neliana* and again longer spur length in *D. capsularis* is selected for, thus, continually driving each other's evolution.

In Agriculture

Pollination management is a branch of agriculture that seeks to protect and enhance present pollinators and often involves the culture and addition of pollinators in monoculture situations, such as commercial fruit orchards. The largest managed pollination event in the world is in Californian almond orchards, where nearly half (about one million hives) of the US honey bees are trucked to the almond orchards each spring. New York's apple crop requires about 30,000 hives; Maine's blueberry crop uses about 50,000 hives each year.

An *Andrena* bee collects pollen among the stamens of a rose. The female carpel structure appears rough and globular to the left. The bee's stash of pollen is on its hind leg.

Bombus ignitus, a popular commercial pollinator in Japan and China

Blueberries being pollinated by bumblebees. Bumblebee hives need to be bought each year as the queens must hibernate (unlike honey bees). They are used nonetheless as they offer advantages with certain fruits as blueberries (such as the fact that they are active even at colder outdoor ambient temperature).

Well-pollinated blackberry blossom begins to develop fruit. Each incipient drupelet has its own stigma and good pollination requires the delivery of many grains of pollen to the flower so that all drupelets develop.

Bees are also brought to commercial plantings of cucumbers, squash, melons, strawberries, and many other crops. Honey bees are not the only managed pollinators: a few other species of bees are also raised as pollinators. The alfalfa leafcutter bee is an important pollinator for alfalfa seed in western United States and Canada. Bumblebees are increasingly raised and used extensively for greenhouse tomatoes and other crops.

The ecological and financial importance of natural pollination by insects to agricultural crops, improving their quality and quantity, becomes more and more appreciated and has given rise to new financial opportunities. The vicinity of a forest or wild grasslands with native pollinators near agricultural crops, such as apples, almonds or coffee can improve their yield by about 20%. The benefits of native pollinators may result in forest owners demanding payment for their contribution in the improved crop results – a simple example of the economic value of ecological services. Farmers can also raise native crops in order to promote native bee pollinator species as shown with *L. vierecki* in Delaware and *L. leucozonium* in southwest Virginia.

The American Institute of Biological Sciences reports that native insect pollination saves the United States agricultural economy nearly an estimated $3.1 billion annually through natural crop production; pollination produces some $40 billion worth of products annually in the United States alone.

Pollination of food crops has become an environmental issue, due to two trends. The trend to monoculture means that greater concentrations of pollinators are needed at bloom time than ever before, yet the area is forage poor or even deadly to bees for the rest of the season. The other trend is the decline of pollinator populations, due to pesticide misuse and overuse, new diseases and parasites of bees, clearcut logging, decline of beekeeping, suburban development, removal of hedges and other habitat from farms, and public concern about bees. Widespread aerial spraying for mosquitoes due to West Nile fears is causing an acceleration of the loss of pollinators.

The US solution to the pollinator shortage, so far, has been for commercial beekeepers to become pollination contractors and to migrate. Just as the combine harvesters follow the wheat harvest from Texas to Manitoba, beekeepers follow the bloom from south to north, to provide pollination for many different crops.

In some situations, farmers or horticulturists may aim to restrict natural pollination to only permit breeding with the preferred individuals plants. This may be achieved through the use of pollination bags.

Improving Pollination in Areas With Suboptimal Bee Densities

In some instances growers' demand for beehives far exceeds the available supply. The number of managed beehives in the US has steadily declined from close to 6 million after WWII, to less than 2.5 million today. In contrast, the area dedicated to growing bee-pollinated crops has grown over 300% in the same time period. Additionally, in the past five years there has been a decline in winter managed beehives, which has reached an unprecedented rate of colony losses at near 30%. At present, there is an enormous demand for beehive rentals that cannot always be met. There is a clear need across the agricultural industry for a management tool to draw pollinators into cultivations and encourage them to preferentially visit and pollinate the flowering crop. By attracting pollinators like honeybees and increasing their foraging behavior, particularly in the center of large plots, we can increase grower returns and optimize yield from their plantings. ISCA Technologies, from Riverside California, created a semiochemical formulation called SPLAT Bloom, that modifies the behavior of honeybees, inciting them to visit flowers in every portion of the field.

Environmental Impacts

Loss of pollinators, also known as Pollinator decline (of which colony collapse disorder is perhaps the most well known) has been noticed in recent years. Observed losses would have significant economic impacts. Possible explanations for pollinator decline include habitat destruction, pesticide, parasitism/diseases, climate change and others, and many researchers believe it is the synergistic effects of these factors which are ultimately detrimental to pollinator populations.

The Structure of Plant–Pollinator Networks

Wild pollinators often visit a large number of plant species and plants are visited by a large number of pollinator species. All these relations together form a network of interactions between plants and pollinators. Surprising similarities were found in the structure of networks consisting out of the interactions between plants and pollinators. This structure was found to be similar in very different ecosystems on different continents, consisting of entirely different species.

The structure of plant-pollinator networks may have large consequences for the way in which pollinator communities respond to increasingly harsh conditions. Mathematical models, examining the consequences of this network structure for the stability of pollinator communities suggest that the specific way in which plant-pollinator networks are organized minimizes competition between pollinators and may even lead to strong indirect facilitation between pollinators when conditions are harsh. This means that pollinator species together can survive under harsh conditions. But it also means that pollinator species collapse simultaneously when conditions pass a critical point. This simultaneous collapse occurs, because pollinator species depend on each other when surviving under difficult conditions.

Such a community-wide collapse, involving many pollinator species, can occur suddenly when increasingly harsh conditions pass a critical point and recovery from such a collapse might not be easy. The improvement in conditions needed for pollinators to recover, could be substantially larger than the improvement needed to return to conditions at which the pollinator community collapsed.

SELF-POLLINATION

Self-pollination is when pollen from the same plant arrives at the stigma of a flower (in flowering plants) or at the ovule (in Gymnosperms). There are two types of self-pollination: In autogamy, pollen is transferred to the stigma of the same flower. In geitonogamy, pollen is transferred from the anther of one flower to the stigma of another flower on the same flowering plant, or from microsporangium to ovule within a single (monoecious)

Gymnosperm. Some plants have mechanisms that ensure autogamy, such as flowers that do not open (cleistogamy), or stamens that move to come into contact with the stigma. The term selfing that is often used as a synonym, is not limited to self-pollination, but also applies to other types of self-fertilization.

One type of automatic self-pollination occurs in the orchid *Ophrys apifera*. One of the two pollinia bends itself towards the stigma.

Occurrence

Few plants self-pollinate without the aid of pollen vectors (such as wind or insects). The mechanism is seen most often in some legumes such as peanuts. In another legume, soybeans, the flowers open and remain receptive to insect cross pollination during the day. If this is not accomplished, the flowers self-pollinate as they are closing. Among other plants that can self-pollinate are many kinds of orchids, peas, sunflowers and tridax. Most of the self-pollinating plants have small, relatively inconspicuous flowers that shed pollen directly onto the stigma, sometimes even before the bud opens. Self-pollinated plants expend less energy in the production of pollinator attractants and can grow in areas where the kinds of insects or other animals that might visit them are absent or very scarce—as in the Arctic or at high elevations.

Self-pollination limits the variety of progeny and may depress plant vigor. However, self-pollination can be advantageous, allowing plants to spread beyond the range of suitable pollinators or produce offspring in areas where pollinator populations have been greatly reduced or are naturally variable.

Pollination can also be accomplished by cross-pollination. Cross-pollination is the transfer of pollen, by wind or animals such as insects and birds, from the anther to the stigma of flowers on separate plants.

Types of Flowers that Self Pollinate

Both hermaphrodite and monoecious species have the potential for self-pollination leading to self-fertilization unless there is a mechanism to avoid it. Eighty percent of all flowering plants are hermaphroditic, meaning they contain both sexes in the same flower, while 5 percent of plant species are monoecious. The remaining 15% would therefore be dioecious (each plant unisexual).

Advantages of Self-Pollination

There are several advantages for self-pollinating flowers. If a given genotype is well-suited for an environment, self-pollination helps to keep this trait stable in the species. Not being dependent on pollinating agents allows self-pollination to occur when bees and wind are nowhere to be found. Self-pollination can be an advantage when the number of flowers are small or widely spaced.

Disadvantages of Self-Pollination

The disadvantages of self-pollination come from a lack of variation that allows no adaptation to the changing environment or potential pathogen attack. Self-pollination can lead to inbreeding depression caused by expression of deleterious recessive mutations, or to the reduced health of the species, due to the breeding of related specimens. This is why many flowers that could potentially self-pollinate have a built-in mechanism to avoid it, or make it second choice at best. Genetic defects in self-pollinating plants cannot be eliminated by genetic recombination and offspring can only avoid inheriting the deleterious attributes through a chance mutation arising in a gamete.

Mixed Mating

About 42% of flowering plants exhibit a mixed mating system in nature. In the most common kind of system, individual plants produce a single flower type and fruits may contain self-pollinated, out-crossed or a mixture of progeny types. Another mixed mating system is referred to as dimorphic cleistogamy. In this system a single plant produces both open, potentially out-crossed and closed, obligately self-pollinated cleistogamous flowers.

Self-Pollinating Species

The evolutionary shift from outcrossing to self-fertilization is one of the most common evolutionary transitions in plants. About 10-15% of flowering plants are predominantly self-fertilizing. A few well-studied examples of self-pollinating species are described below.

Paphiopedilum parishii

Self-pollination in the slipper orchid *Paphiopedilum parishii* occurs when the anther changes from a solid to a liquid state and directly contacts the stigma surface without the aid of any pollinating agent.

Holcoglossum amesianum

The tree-living orchid *Holcoglossum amesianum* has a type of self-pollination mechanism in which the bisexual flower turns its anther against gravity through 360° in order to insert pollen into its own stigma cavity---without the aid of any pollinating agent or medium. This type of self-pollination appears to be an adaptation to the windless, drought conditions that are present when flowering occurs, at a time when insects are scarce. Without pollinators for outcrossing, the necessity of ensuring reproductive success appears to outweigh potential adverse effects of inbreeding. Such an adaptation may be widespread among species in similar environments.

Caulokaempferia coenobialis

In the Chinese herb *Caulokaempferia coenobialis* a film of pollen is transported from the anther (pollen sacs) by an oily emulsion that slides sideways along the flower's style and into the individual's own stigma. The lateral flow of the film of pollen along the style appears to be due solely to the spreading properties of the oily emulsion and not to gravity. This strategy may have evolved to cope with a scarcity of pollinators in the extremely shady and humid habitats of *C. coenobialis*.

Capsella rubella

Capsella rubella (Red Shepard's purse) is a self-pollinating species that became self-compatible 50,000 to 100,000 years ago, indicating that self-pollination is an evolutionary adaptation that can persist over many generations. Its out-crossing progenitor was identified as *Capsella grandiflora*.

Arabidopsis thaliana

Arabidopsis thaliana is a predominantly self-pollinating plant with an out-crossing rate in the wild estimated at less than 0.3%. A study suggested that self-pollination evolved roughly a million years ago or more.

Possible Long-Term Benefit of Meiosis

Meiosis followed by self-pollination produces little overall genetic variation. This raises the question of how meiosis in self-pollinating plants is adaptively maintained over extended periods [i.e. for roughly a million years or more, as in the case of *A.thaliana*] in preference to a less complicated and less costly asexual ameiotic process for producing progeny. An adaptive benefit of meiosis that may explain its long-term maintenance in self-pollinating plants is efficient recombinational repair of DNA damage. This benefit can be realized at each generation (even when genetic variation is not produced).

REFERENCES

1. Fritsch, Felix Eugene; Salisbury, Edward James (1920). "An introduction to the structure and reproduction of plants". G. Bell.
2. Mauseth, James D. Botany: An Introduction to Plant Biology. Publisher: Jones & Bartlett, 2008 ISBN 978-0-7637-5345-0
3. Raghavan, Valayamghat (1997). Molecular Embryology of Flowering Plants. Cambridge University Press. pp. 210–216. ISBN 978-0-521-55246-2.
4. Campbell, Neil A.; Reece, Jane B. (2002). Biology (6th edition). Pearson Education. pp. 600–612. ISBN 978-0-201-75054-6.
5. Runions, C. John; Owens, John N. (1999). "Sexual reproduction of interior spruce (Pinaceae). I. Pollen germination to archegonial maturation". International Journal of Plant Sciences. 160 (4): 631–640. doi:10.1086/314170.
6. Abrol, Dharam P. (2012). "Non Bee Pollinators-Plant Interaction". Pollination Biology. Chapter 9: 265–310. doi:10.1007/978-94-007-1942-2_9.
7. Faegri, K.; Van Der Pijl, L. (2013). Principles of Pollination Ecology. Elsevier. pp. 34–40. ISBN 978-1-4832-9303-5.
8. "Types of Pollination, Pollinators and Terminology". CropsReview.Com. Retrieved 2015-10-20.
9. "Pollinator Syndromes". fs.fed.us. Retrieved 2015-10-20.
10. "First ever record of insect pollination from 100 million years ago". ScienceDaily. Retrieved 2015-10-20.
11. Rodríguez-Gironés, Miguel A.; Santamaría, Luis (2004). "Why are so many bird flowers red?". PLoS Biology. 2 (10): e306. doi:10.1371/journal.pbio.0020350. PMC 521733Freely accessible. PMID 15486585.
12. Hill, P.S.M.; Wells, P.H.; Wells, H. (1997). "Spontaneous flower constancy and learning in honey bees as a function of colour". Animal Behaviour. 54: 615–627. doi:10.1006/anbe.1996.0467.
13. Stout, J.C.; Allen, J.A.; Goulson, D. (1998). "The influence of relative plant density and floral morphological complexity on the behaviour of bumblebees". Oecologia. 117: 543–550. doi:10.1007/s004420050691.
14. Chittka, L.; Gumbert, A.; Kunze, J. (1997). "Foraging dynamics of bumble bees: correlates of movement within and between plant species". Behavioral Ecology. 8 (3): 239–249. doi:10.1093/beheco/8.3.239.
15. Goulson, D.; Ollerton, J.; Sluman, C. (1997). "Foraging strategies in the small skipper butterfly, Thymelicus flavus: when to switch?". Animal Behaviour. 53: 1009–1016. doi:10.1006/anbe.1996.0390.
16. Harder, L. D., N.M. Williams, C.Y. Jordan, and W.A. Nelson. "The effects of Floral design and display on pollinator economics and pollen dispersal". 297–317.
17. Editors, L. Chittka and J.D. Thomson. Cognitive Ecology of Pollination: Animal Behavior and Floral Evolution. 2001. Cambridge University Press.
18. Chittka, L.; Thomson, J.D.; Waser, N.M. (1999). "Flower constancy, insect

psychology, and plant evolution". Naturwissenschaften. 86: 361–177. doi:10.1007/s001140050636.

19. Cronk, J. K.; Fennessy, M. Siobhan (2001). Wetland plants: biology and ecology. Boca Raton, Fla.: Lewis Publishers. p. 166. ISBN 1-56670-372-7.
20. Glover, Beverly J. (2007). Understanding flowers and flowering: an integrated approach. Oxford University Press. p. 127. ISBN 0-19-856596-8.
21. New Living Science: Biology for Class 9. Ratna Sagar. pp. 56–61. ISBN 978-81-8332-565-3.
22. Culley, Theresa M.; Klooster, Matthew R. (2007). "The cleistogamous breeding system: a review of its frequency, evolution, and ecology in angiosperms". The Botanical Review. 73: 1–30. doi:10.1663/0006-8101(2007)73[1:TCBSAR]2.0.CO;2.
23. Baskin, Carol C.; Baskin, Jerry M. (2001). Seeds: Ecology, Biogeography, and Evolution of Dormancy and Germination. Elsevier. p. 215. ISBN 978-0-12-080263-0.
24. Igic B, Kohn JR (May 2006). "The distribution of plant mating systems: study bias against obligately outcrossing species". Evolution. 60 (5): 1098–103. doi:10.1554/05-383.1. PMID 16817548.
25. Goodwillie C, Kalisz S, Eckert CG (2005) The evolutionary enigma of mixed mating systems in plants: Occurrence, theoretical explanations, and empirical evidence. Annu. Rev. Ecol. Evol. Syst. 36: 47–79. doi:10.1146/annurev.ecolsys.36.091704. 175539
26. Potts, Brad, Gore Peter. Reproductive Biology and Controlled Pollination of Eucalyptus. School of Plant Science, University of Tasmania 1995
27. Dressler, Robert L. (1968-03 01). "Pollination by Euglossine Bees". Evolution. 22 (1): 202–210. doi:10.2307/2406664. JSTOR 2406664.
28. Olesen, Jens M. & Valido, Alfredo. Lizards as pollinators and seed dispersers: an island phenomenon. TRENDS in Ecology and Evolution Vol.18 No.4 April 2003
29. GODÍNEZ-ÁLVAREZ HÉCTOR Pollination and seed dispersal by lizards. Revista Chilena de Historia Natural 77: 569–577, 2004
30. Wiens, Delbert; Rourke, John P.; Casper, Brenda B.; Rickart, Eric A.; LaPine, T.R.; Peterson,J.; Channing, A: Nonflying Mammal Pollination of Southern African Proteas. Annals of the Missouri Botanical Garden Vol 70, number 1, 1983
31. P.A. Fleming & S.W. Nicolson. Arthropod fauna of mammal-pollinated Protea humiflora: ants as an attractant for insectivore pollinators? African Entomology 11(1): 9–14 (2003)
32. http://protea.worldonline.co.za/p52prhumi.htm
33. Goldingay, R. L.; Carthew, S. M.; Whelan, R. J. (May 1991). "The Importance of Non-Flying Mammals in Pollination". Oikos. Wiley-Blackwell. pp. 79–87. JSTOR 3545409.
34. Sühs, R.B.; Somavilla, A.; Putzke, J.; Köhle, A. (2009). "Pollen vector wasps (Hymenoptera, Vespidae) of Schinus terebinthifolius Raddi (Anacardiaceae)". Brazilian Journal of Biosciences. Santa Cruz do Sul, RS, Brazil. pp. 138–143.

35. W. Scott Armbruster (2012). "3". In Patiny, Sébastien. Evolution of Plant-Pollinator Relationships. Cambridge, UK: Cambridge University Press. pp. 45–67.
36. Cardinal, Sophie; Danforth, Bryan N. (2013). "Bees diversified in the age of eudicots". Proceedings of the Royal Society. 280: 20122686. doi:10.1098/rspb.2012.2686.
37. Steiner, Kim E.; Whitehead, V. B. (September 1990). "Pollinator adaptation to oil-secreting flowers—Rediviva and Diascia". Evolution. 44 (6): 1701–1707. doi:10.2307/2409348.
38. Shao, Z.-Y.; Mao, H.-X.; Fu, W.-J.; Ono, M.; Wang, D.-S.; Bonizzoni, M.; Zhang, Y.-P. (2004-01-01). "Genetic Structure of Asian Populations of Bombus ignitus (Hymenoptera: Apidae)". Journal of Heredity. 95 (1): 46–52. doi:10.1093/jhered/esh008. ISSN 0022-1503. PMID 14757729.
39. Kuehn, F. Coordinator. (2015). Farming for native bees. World Wide Web electronic publication. Retrieved from . (Accessed: September 22, 2015).
40. Adamson, Nancy Lee. An Assessment of Non-Apis Bees as Fruit and Vegetable Crop Pollinators in Southwest Virginia. Diss. 2011. Web. 15 Oct. 2015.
41. BioScience, April 2006, Vol. 56 No. 4, pp. 315–317
42. "US Forest Department: Pollinator Factsheet" (PDF). Retrieved 2014-04-18.
43. Biesmeijer, JC; Roberts, SPM; Reemer, M; Ohlemuller, R; Edwards, M; Peeters, T; et al. (2006). ", Parallel declines in pollinators and insect-pollinated plants in Britain and the Netherlands". Science. 313: 351–354. doi:10.1126/science.1127863.
44. Cox-Foster, DL; Conlan, S; Holmes, EC; Palacios, G; Evans, JD; Moran, NA; et al. (2007). ", A metagenomic survey of microbes in honey bee colony collapse disorder". Science. 318: 283–287. doi:10.1126/science.1146498. PMID 17823314.
45. Woteki, Catherine (August 16, 2013). "The road to pollinator health". Science. 341 (6147): 695. doi:10.1126/science.1244271.
46. "EFSA Press Release: EFSA identifies risks to bees from neonicotinoids". Efsa.europa.eu. Retrieved 2014-04-18.
47. "ISCA Technologies: A Leader of Innovative Pest Management Tools and Solutions". Iscatech.com. Retrieved 2014-04-18.
48. David Ward Roubik. Ups and Downs in Pollinator Populations: When is there a Decline?
49. Bascompte, J.; Jordano, P.; Melián, C. J.; Olesen, J. M. (2003). "The nested assembly of plant–animal mutualistic networks". Proceedings of the National Academy of Sciences. 100 (16): 9383–9387. doi:10.1073/pnas.1633576100.
50. Bastolla, U.; Fortuna, M. A.; Pascual-García, A.; Ferrera, A.; Luque, B.; Bascompte, J. (2009). "The architecture of mutualistic networks minimizes competition and increases biodiversity". Nature. 458 (7241): 1018–1020. doi:10.1038/nature07950.
51. Lever, J. J.; Nes, E. H.; Scheffer, M.; Bascompte, J. (2014). "The sudden collapse of pollinator communities". Ecology Letters. 17 (3): 350–359. doi:10.1111/ele.12236.

52. Grossenbacher D, Briscoe Runquist R, Goldberg EE, Brandvain Y (2015). "Geographic range size is predicted by plant mating system". Ecol. Lett. 18 (7): 706–13. doi:10.1111/ele.12449. PMID 25980327.

53. Charlesworth D, Willis JH (2009). "The genetics of inbreeding depression". Nat. Rev. Genet. 10 (11): 783–96. doi:10.1038/nrg2664. PMID 19834483.

54. Goodwillie C, Kalisz S, Eckert CG (2005) The evolutionary enigma of mixed mating systems in plants: Occurrence, theoretical explanations, and empirical evidence. Annu. Rev. Ecol. Evol. Syst. 36: 47-79. doi:10.1146/annurev.ecolsys.36.091704. 175539

55. Munguía-Rosas MA, Campos-Navarrete MJ, Parra-Tabla V (2013). "The effect of pollen source vs. flower type on progeny performance and seed predation under contrasting light environments in a cleistogamous herb". PLoS ONE. 8 (11): e80934. doi:10.1371/journal.pone.0080934. PMC 3829907Freely accessible. PMID 24260515.

56. Wright SI, Kalisz S, Slotte T (June 2013). "Evolutionary consequences of self-fertilization in plants". Proc. Biol. Sci. 280 (1760): 20130133. doi:10.1098/rspb.2013.0133. PMC 3652455Freely accessible. PMID 23595268.

57. Chen LJ, Liu KW, Xiao XJ, Tsai WC, Hsiao YY, Huang J, Liu ZJ (2012). "The anther steps onto the stigma for self-fertilization in a slipper orchid". PLoS ONE. 7 (5): e37478. doi:10.1371/journal.pone.0037478. PMC 3359306Freely accessible. PMID 22649529.

58. Liu KW, Liu ZJ, Huang L, Li LQ, Chen LJ, Tang GD (June 2006). "Pollination: self-fertilization strategy in an orchid". Nature. 441 (7096): 945–6. doi:10.1038/441945a. PMID 16791185.

59. Wang Y, Zhang D, Renner SS, Chen Z (September 2004). "Botany: a new self-pollination mechanism". Nature. 431 (7004): 39–40. doi:10.1038/431039b. PMID 15343325.

60. Brandvain Y, Slotte T, Hazzouri KM, Wright SI, Coop G (2013). "Genomic identification of founding haplotypes reveals the history of the selfing species Capsella rubella". PLoS Genet. 9 (9): e1003754. doi:10.1371/journal.pgen.1003754. PMC 3772084Freely accessible. PMID 24068948.

61. Slotte T, Hazzouri KM, Ågren JA, Koenig D, Maumus F, Guo YL, Steige K, Platts AE, Escobar JS, Newman LK, Wang W, Mandáková T, Vello E, Smith LM, Henz SR, Steffen J, Takuno S, Brandvain Y, Coop G, Andolfatto P, Hu TT, Blanchette M, Clark RM, Quesneville H, Nordborg M, Gaut BS, Lysak MA, Jenkins J, Grimwood J, Chapman J, Prochnik S, Shu S, Rokhsar D, Schmutz J, Weigel D, Wright SI (July 2013). "The Capsella rubella genome and the genomic consequences of rapid mating system evolution". Nat. Genet. 45 (7): 831–5. doi:10.1038/ng.2669. PMID 23749190.

62. Population genetic structure and outcrossing rate of Arabidopsis thaliana (L.) Heynh. Abbott RJ, Gomes MF. Heredity 1989 62:411-418

63. Tang C, Toomajian C, Sherman-Broyles S, Plagnol V, Guo YL, Hu TT, Clark RM, Nasrallah JB, Weigel D, Nordborg M (August 2007). "The evolution of

selfing in Arabidopsis thaliana". Science. 317 (5841): 1070–2. doi:10.1126/science.1143153. PMID 17656687.

64. Bernstein H, Hopf FA, Michod RE. (1987). The molecular basis of the evolution of sex. Adv Genet 24:323-70. Review. PMID 3324702
65. Harris Bernstein, Carol Bernstein and Richard E. Michod (2011). Meiosis as an Evolutionary Adaptation for DNA Repair. Chapter 19 in DNA Repair. Inna Kruman editor. InTech Open Publisher. DOI: 10.5772/25117 http://www.intechopen.com/books/dna-repair/meiosis-as-an-evolutionary-adaptation-for-dna-repair

CHAPTER 7

Breeding for Drought Stress Tolerance

Chapter Contents

- **Introduction**
- **Drought Stress in Crop Plants**
- **Drought Stress Affects Plant Physiology**
- **Mechanism of Drought Stress Resistance**
- **Breeding for Heat Stress Tolerance**

Plant responses to drought stress are very complex as stress itself involves various climatic, soil and agronomic factors, frequently complicated by substantial variation in timing of occurrence, duration and intensity. The general complexity of drought is often aggravated under rainfed conditions in marginal areas by erratic and unpredictable rainfall, and by the occurrence of high temperatures, high levels of solar radiation, and poor soil characteristics. This high variability in the nature of drought stress and an insufficient understanding of its complexity, have made it difficult to identify specific physiological traits required for improved crop performance under drought, consequently limiting plant breeding efforts to enhance crop drought tolerance. It is highly probable that optimal drought-adaptation requires the combination of several morphological, physiological, and phenological processes, which depend on a multitude of genes and varies within each target environment.

In addition to its direct effects on yield, drought can also reduce the potential beneficial effects of improved crop management practices such as fertilizer application or pest and disease management. Given the increasing scarcity of water resources, and competition for them, irrigation is not a practical option for alleviating drought in most of the rainfed areas. Drought management strategies therefore need to focus on maximizing extraction of available soil moisture and the efficiency of its use in crop establishment, growth, biomass and grain yield. However, agronomic and genetic options that do not involve external inputs of irrigation can only partially alleviate drought effects, because yield will always be lower than what can be achieved with irrigation.

INTRODUCTION

Breeding for drought stress tolerance is the process of breeding plants with the goal of reducing the impact of drought on plant growth.

DROUGHT STRESS IN CROP PLANTS

In nature or crop fields, water is often the most limiting factor for plant growth. If plants do not receive adequate rainfall or irrigation, the resulting drought stress can reduce growth more than all other environmental stresses combined.

Drought can be defined as the absence of rainfall or irrigation for a period of time sufficient to deplete soil moisture and injure plants. Drought stress results when water loss from the plant exceeds the ability of the plant's roots to absorb water and when the plant's water content is reduced enough to interfere with normal plant processes.

Drought Stress is a Global Phenomenon

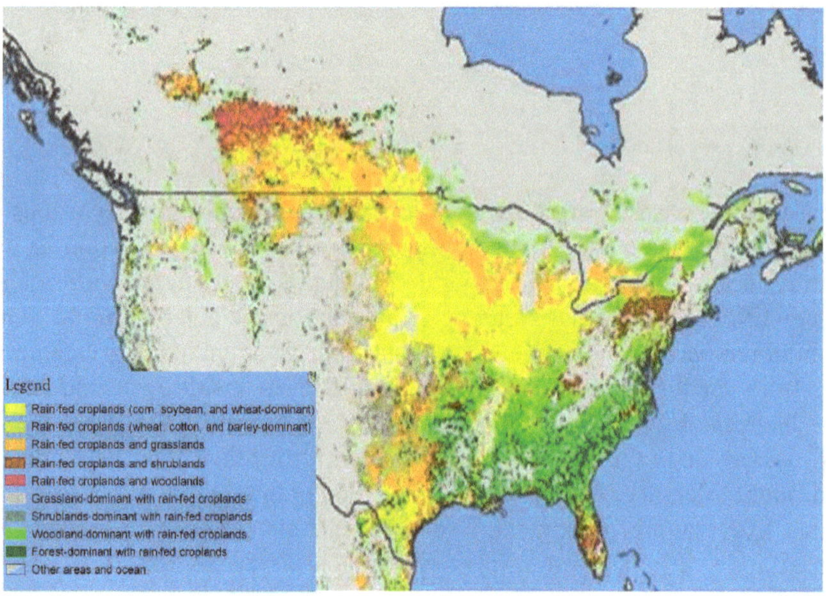

Rain fed areas of USA, for details - Biradar et al.(2009) International J Applied Earth Observation and Geoinformation 11: 114-129.

About 15 million km2 of the land surface is covered by cropland (), and about 16% of this area is equipped for irrigation (Siebert *et al.* 2005). Thus in many parts of the world including the United States, plants may frequently encounter drought stress. Rainfall is very seasonal and periodic drought occurs regularly. The effect of drought is more prominent in sandy soils with low water holding capacity. On such soils some plants may experience drought stress after only a few days without water.

During the last century, the rate of increase in `blue' water withdrawal (from rivers, lakes, and aquifers) for irrigation and other purposes was higher than the growth rate of the world population (Shiklomanov 1998). Country-wise maps of irrigated areas are available.

Drought Stress and Future Challenges to Crop Production

Moisture deficits is a significant challenge to the future crop production. Severe drought in parts of the U.S., Australia, and Africa in recent years drastically reduced crop yields and disrupted regional economies. Even in average years, however, many agricultural regions, including the U.S. Great Plains, suffer from chronic moisture deficits. Cereal crops typically attain only about 25% of their potential yield due to the effects of environmental stress, with moisture stress the most important cause. Two major trends will likely increase the frequency and severity of crop moisture deficits:

(1) Global Climate Change.

Higher temperatures are likely to increase crop water use due to increased transpiration. A warmer atmosphere will also speed up melting of mountain snowpack, resulting in less water available for irrigation. More extreme weather patterns will increase the frequency of drought in some regions.

(2) Competing Uses for Limited Water Supplies.

Increased demand from municipal and industrial users will further reduce the amount of water available for irrigated crops.

Although changes in tillage and irrigation practices can improve production by conserving water, enhancing the genetic tolerance of crops to drought stress is considered an essential strategy for addressing moisture deficits.

DROUGHT STRESS AFFECTS PLANT PHYSIOLOGY

A plant responds to a lack of water by halting growth and reducing photosynthesis and other plant processes in order to reduce water use. As water loss progresses, leaves of some species may appear to change color—usually to blue-green. Foliage begins to wilt and, if the plant is not irrigated, leaves will fall off and the plant will eventually die. Drought lowers the water potential of a plant's root and upon extended exposure, abscisic acid is accumulated and eventually stomatal closure occurs. This reduces a plant's leaf relative water content. The time required for drought stress to occur depends on the water-holding capacity of the soil, environmental conditions, stage of plant growth, and plant species. Plants growing in sandy soils with low water-holding capacity are more susceptible to drought stress than plants growing in clay soils. A limited root system will accelerate the rate at which drought stress develops. A root system may be limited by the presence of competing root systems, by site conditions such as compacted soils or high water tables, or by container size (if growing in a container). A plant with a large mass of leaves in relation to the root system is prone to drought stress because the leaves may lose water faster than the roots can supply it. Newly installed plants and poorly established plants may be especially susceptible to drought stress because of the limited root system or the large mass of stems and leaves in comparison to roots.

Drought Stress Interaction with Other Stress Factors

Aside from the moisture content of the soil, environmental conditions of high light intensity, high temperature, low relative humidity and high wind speed will significantly increase plant water loss. The prior environment of a plant also can influence the development of drought stress. A plant that has been drought stressed previously and has recovered may become more drought resistant. Also, a plant that was well-watered prior to drought will usually survive drought better than a continuously drought-stressed plant.

MECHANISM OF DROUGHT STRESS RESISTANCE

Degree of Resistance to drought depend upon crops. Generally three strategies can help a crop to mitigate the effect of drought stress:

- drought escape
- drought avoidance
- drought tolerance

A proper timing of lifecycle, resulting in the completion of the most sensitive developmental stages while water is abundant, is considered to be a drought escape strategy. Avoiding water-deficit stress with a root system capable of extracting water from deep soil layers, or by reducing evapotranspiration without affecting yields, is considered as drought avoidance. Mechanisms such as osmotic adjustment (OA) whereby a plant maintains cell turgor pressure under reduced soil water potential are categorized as drought tolerance mechanisms. Drought avoidance mechanisms can be expressed even in the absence of stress and are then considered constitutive. Drought tolerance mechanisms are the result of a response triggered by drought stress itself and are therefore considered adaptative. When the stress is terminal and predictable, drought escape through the use of shorter duration varieties is often the preferable method of improving yield potential. Drought avoidance and tolerance mechanisms are required in situations where the timing of drought is mostly unpredictable.

Drought tolerance mechanism is genetically controlled and genes or QTL responsible for drought tolerance have been discovered in several crops which opens avenue for molecular breeding for drought tolerance.

Drought Tolerance Traits

Tolerance to drought is a quantitative trait, with a complex phenotype, often confounded by plant phenology. Breeding for drought tolerance is further complicated since several types of abiotic stress, such as high temperatures, high irradiance, and nutrient toxicities or deficiencies can challenge crop plants simultaneously.

(a) Osmotic adjustment When a plant is exposed to water deficit, it may accumulate a variety of osmotically active compounds such as amino acids and sugars, resulting in a lowering of the osmotic potential. Examples of amino acids that may be up-regulated are proline and glycine betaine. This is termed osmotic adjustment and enables the plant to take up water, maintain turgor and survive longer.

(b) Cell membrane stability The ability to survive dehydration is influenced by a cell's ability to survive at reduced water content. This can be considered complementary to OA because both traits will help maintain leaf growth (or prevent leaf death) during drought. Crop varieties differ in dehydration tolerance and an important factor for such differences is the capacity of the cell membrane to prevent electrolyte leakage at decreasing water content, or "cell membrane stability (CMS)". The maintenance of membrane function is assumed to mean that cell activity is also maintained. Measurements of CMS have been used in different crops and are known to be correlated with yields under high temperature and possibly under drought stress.

(c) Epicuticular wax In sorghum (Sorghum bicolor L. Moench), drought resistance is a trait that is highly correlated with the thickness of the epicuticular wax layer. Experiments have demonstrated that rice varieties with a thick cuticle layer retain their leaf turgor for longer periods of time after the onset of a water-stress.

(d) Partitioning and stem reserve mobilization

As photosynthesis becomes inhibited by drought, the grain filling process becomes increasingly reliant on stem reserve utilization.72 Numerous studies have reported that stem reserve mobilization capacity is related to yield under water-stress in wheat. In rice, a few studies also indicated that this mechanism maintains grain yield under water stress at the grain filling stage. This drought tolerance mechanism is stimulated by a decrease in gibberellic acid concentration and an increase in abscisic acid concentration.

(e) Manupulation and Stability of flowering processes

(g) Seedling drought traits For emergence from deep sowing (to exploit dry upper soil), this is practiced to help seedlings reach the receding moisture profile, and to avoid high soil surface temperatures which inhibit germination. Screening at these stage provides practical advantages, specially when managing large amount of germplasms.

Crop Ideotype for Drought Tolerance

Usually ideotypes are developed to create an ideal plant variety. The following traits constitutes ideotype of wheat by CIMMYT.

1) Large seed size. Helps emergence, early ground cover, and initial biomass.
2) Long coleoptiles. For emergence from deep sowing
3) Early ground cover. Thinner, wider leaves (i.e., with a relatively low specific leaf weight) and a more prostrate growth habit help to increase ground cover, thus conserving soil moisture and potentially increasing radiation use efficiency.
4) High pre-anthesis biomass.
5) Good capacity for stem reserves and remobilization
6) High spike photosynthetic capacity
7) High RLWC/Gs/CTD during grain filling to indicate ability to extract water
8) Osmotic adjustment
(9) Accumulation of ABA.

The benefit of ABA accumulation under drought has been demonstrated (Innes et al. 1984). It appears to pre-adapt plants to stress by reducing stomatal conductance, rates of cell division, organ size,and increasing development rate. However, high ABA can also result in sterility problems since high ABA levels may abort developing florets

10) Heat Tolerance. The contribution of heat tolerance to performance under moisture stress needs to be quantified, but it is relatively easy to screen for (Reynolds et al. 1998).
11) Leaf anatomy: waxiness, pubescence, rolling, thickness, posture. These traits decrease radiation load to the leaf surface. Benefits include a lower evapotranspiration rate and reduced risk of irreversible photo-inhibition. However, they may also be associated with reduce radiation use efficiency, which would reduce yield under more favorable conditions.
12) High tiller survival.

Comparison of old and new varieties have shown that under drought older varieties over-produce tillers many of which fail to set grain while modern drought tolerant lines produce fewer tillers most of which survive.

13) Stay-green.

The trait may indicate the presence of drought avoidance mechanisms, but probably does not contribute to yield per se if there is no water left in the soil profile by the end of the cycle to support leaf gas exchange. It may be detrimental if it indicates lack of ability to remobilize stem reserves. However, research in sorghum has indicated that staygreen is associated with higher leaf chlorophyll content at all stages of development and both were associated with improved yield and transpiration efficiency under drought.

Combination Phenomics: Overall Health of Crops

The concept of combination phenomics comes from the idea that two or more stresses have common physiological effects or common traits - which are an indicator of overall plant health. Similar analogy in human medical terms is high blood pressure or high body temperature or high white blood cells in body is an indicator of health problems and thus we can select healthy people from unhealthy using such a measure. As both abiotic and abiotic stresses can result in similar physiological consequence, tolerant plant can be separated from sensitive plants. Some imaging or infrared measuring techniques can help to speed the process for breeding process. For example spot blotch intensity and canopy temperature depression can be monitored with canopy temperature depression.

Molecular Breeding for Drought Tolerance

Recent research breakthroughs in biotechnology have revived interest in targeted drought tolerance breeding and use of new genomics tools to enhance crop water productivity. Marker-assisted breeding is revolutionizing the improvement of temperate field crops and will have similar impacts on breeding of tropical crops. Other molecular breeding tool include development of genetically modified crops that can tolerate plant stress. As a complement to the recent rapid progress in genomics, a better understanding of physiological mechanisms of drought response will also contribute to the progress of genetic enhancement of crop drought tolerance. It is now well accepted that the complexity of the drought syndrome can only be tackled with a holistic approach that integrates physiological dissection of crop drought avoidance and tolerance traits using molecular genetic tools such as MAS, microarrays and transgenic crops, with agronomic practices that lead to better conservation and utilization of soil moisture, and better matching of crop genotypes with the environment.

BREEDING FOR HEAT STRESS TOLERANCE

Plant breeding is process of development of new cultivars. Plant breeding involves development of varieties for different environmental conditions – some of them are not favorable. Among them, heat stress is one of such factor that reduces the production and quality significantly. So breeding against heat is a very important criterion for breeding for current as well as future environments produced by global climate change (e.g. global warming).

Breeding for Heat Stress Tolerance in Plants

Heat stress due to increased temperature is a very important problem globally. Occasional or prolonged high temperatures cause different morpho-anatomical, physiological and biochemical changes in plants. The ultimate effect is on plant growth as well as development and reduced yield and quality. Breeding for heat stress tolerance can be mitigated by breeding plant varieties that have improved levels of thermo-tolerance using different conventional or advanced genetic tools.

What is Heat Stress Tolerance

Heat stress is defined as increased temperature level sufficient to cause irreversible damage to plant growth and development. Generally a temperature rise, above usually 10 to 15 °C above ambient, can be considered heat shock or heat stress. Heat tolerance is broadly defined as the ability of the plant tolerate heat – means that grow and produce economic yield under high temperatures.

Significance: Current and Future - Global Warming

Heat stress is a serious threat to crop production globally (Hall, 2001, 1992). Global warming is particularly consequence of increased level of green house gases such as CO_2, methane, chlorofluorocarbons and nitrous oxides. The Intergovernmental Panel on Climatic Change (IPCC) has predicted a rise of 0.3 °C per decade (Jones et al., 1999) reaching to approximately 1 and 3 °C above the present value by 2025 and 2100 AD, respectively.

Physiological Consequence of Heat Stress

At very high temperatures cause severe cellular injury and cell death may occur within short time, thus leading to a catastrophic collapse of cellular organization (Schoffl et al., 1999). However, under moderately high temperatures, the injury can only occur after longer exposure to such a temperature however the plant efficiency can be severely affected. High temperature directly affect injuries such as protein denaturation and aggregation, and increased fluidity of membrane lipids. Other indirect or slower heat injuries involve inactivation of enzymes in chloroplast and mitochondria, protein degradation, inhibition of protein synthesis, and loss of membrane integrity. Heat stress associated injuries ultimately lead to starvation, inhibition of growth, reduced ion flux, production of toxic compounds and production of reactive oxygen species (ROS). Immediately after exposure to high temperature stress-related proteins are expressed as stress defense strategy of the cell. Expression of heat shock proteins (HSPs), protein with 10 to 200 kDa, is supposed to be involved in signal transduction during heat stress. In many species it has been demonstrated that HSPs results in improved physiological phenomena such as photosynthesis, assimilate partitioning, water and nutrient use efficiency, and membrane stability.

Studies have found tremendous variation within and between species, thus this will help to breed heat tolerance for future environment. Some of attempts to develop heat-tolerant genotypes are successful. (Ehlers and Hall, 1998; Camejo et al., 2005)

Traits Associated with Heat Stress Tolerance

Different physiological mechanisms may contribute to heat tolerance in the field—for example, heat tolerant metabolism as indicated by higher photosynthetic rates, stay-green, and membrane thermo-stability, or heat avoidance as indicated by canopy temperature depression. Several physiological and morphological traits have been evaluated for heat tolerance - Canopy temperature, leaf chlorophyll, stay green, leaf conductance, spike number, biomass, and flowering date.

(a) Canopy Temperature Depression (CTD)

CTD has shown clear association with yield in warm environments shows it association with heat stress tolerance. CTD shows high genetic correlation with yield and high values of proportion of direct response to selection (Reynolds et al., 1998) indicating that the trait is heritable and therefore amenable to early generation selection. Since an integrated CTD value can be measured almost instantaneously on scores of plants in a small breeding plot (thus reducing error normally associated with traits measured on individual plants), work has been conducted to evaluate its potential as an indirect selection criterion for genetic gains in yield. CTD is affected by many physiological factors, which makes it a powerful.

(b) Stomatal Conductance

Canopy temperature depression is highly suitable for selecting physiologically superior lines in warm, low relative humidity environments where high evaporative demand leads to leaf cooling of up to 10 °C below ambient temperatures. This permits differences among genotypes to be detected relatively easily using infrared thermometry. However, such differences cannot be detected in high relative humidity environments because the effect of evaporative cooling of leaves is negligible. Nonetheless, leaves maintain their stomata open to permit the uptake of CO2, and differences in the rate of CO2 fixation may lead to differences in leaf conductance that can be measured using a porometer. Porometry can be used to screen individual plants. The heritability of stomatal conductance is reasonably high, with reported values typically in the range of 0.5 to 0.8. Plants can be assessed for leaf conductance using a viscous flow porometer that is available on the market (Thermoline and CSIRO, Australia). This instrument can give a relative measure of stomatal conductance in a few seconds, making it possible to identify physiologically superior genotypes from within bulks.

(C) Membrane Thermostability

Although resistance to high temperatures involves several complex tolerance and avoidance mechanisms, the membrane is thought to be a site of primary physiological injury by heat, and measurement of solute leakage from tissue can be used to estimate damage to membranes. Since membrane thermostability is reasonably heritable (Fokar et al., 1998) and shows high genetic correlation with yield.

(D) Chlorophyll Fluorescence

Chlorophyll fluorescence, an indication of the fate of excitation energy in the photosynthetic apparatus, has been used indicator for heat stress tolerance.

(E) Chlorophyll Content and Stay green

Chlorophyll content and stay green traits have been found to be associated with heat stress tolerance.,. Xu et al. (2000) identified three QTLs for chlorophyll content (Chl1,

Chl2, and Chl3) (coincided with three stay-green QTL regions (Stg1, Stg2, and Stg3)) were identified in Sorghum. The Stg1 and Stg2 regions also contain the genes for key photosynthetic enzymes, heat shock proteins, and an abscisic acid (ABA) responsive gene.

(F) Photosynthesis

Decliend photosynthesis is suggested as measure of heat stress sensitivity in plants.

(G) Stem Reserve Remobilization'

Combination Breeding and Physiological Breeding

The physiological-trait-based breeding approach has merit over breeding for yield *per se* because it increases the probability of crosses resulting in additive gene action. The concept of combination phenomics comes from the idea that two or more stress have common physiological effect or common traits - which is an indicator of overall plant health. Similar analogy in human medical terms is high blood pressure or high body temperature or high white blood cells in body is an indicator of health problems and thus we can select healthy people from unhealthy using such a measure. As both abiotic and abiotic stresses can result in similar physiological consequence, tolerant plant can be separated from sensitive plants. Some imaging or infrared measuring techniques can help to speed the process for breeding process. For example, spot blotch intensity and canopy temperature depression can be monitored with canopy temperature depression.

REFERENCES

1. Ramankutty N, Evan A T, Monfreda C and Foley J A 2008 Farming the planet. Part 1: the geographic distribution of global agricultural lands in the year 2000 Glob. Biogeochem. Cycles 22 GB1003
2. Siebert, S; Döll, P; Hoogeveen, J; Faures, J M; Frenken, K; Feick, S (2005). "Development and validation of the global map of irrigation areas". Hydrol. Earth Syst. Sci. 9: 535–47. doi:10.5194/hess-9-535-2005.
3. Shiklomanov I A 1998 World Water Resources—A New Appraisal and Assessment for the 21st Century
4. Stefan Siebert, Petra Döll, Sebastian Feick, Jippe Hoogeveen and Karen Frenken (2007) Global Map of Irrigation Areas version 4.0.1. Johann Wolfgang Goethe University, Frankfurt am Main, Germany / Food and Agriculture Organization of the United Nations, Rome, Italy
5. ftp://ftp.fao.org/agl/aglw/aquastat/GMIAv401hires.pdf
6. http://www.sciencedirect.com/cache/MiamiImageURL/1-s2.0-S0303243408000834-gr7_lrg.jpg/0?wchp=dGLzVlk-zSkzk
7. Ogbaga, Chukwuma C.; Stepien, Piotr; Johnson, Giles N. (October 2014). "Sorghum (Sorghum bicolor) varieties adopt strongly contrasting strategies in

response to drought". Physiologia Plantarum. 152 (2): 389–401. doi:10.1111/ppl.12196.

8. Rosyara, U.R., A. A. Ghimire, S. Subedi, and Ram C. Sharma. 2008.Variation in south Asian wheat germplasm for seedling drought tolerance traits. Plant Genetic Resources 6: 88-93
9. Richards, R.A. (1996). "Defining selection criteria to improve yield under drought". Plant Growth Regulation. 20: 157–166. doi:10.1007/bf00024012.
10. Innes, P., R.D. Blackwell, and S.A. Quarrie. 1984. Some effects of genetic variation in drought-induced abscisic acid accumulation on the yield and water-use of spring J. Agric. Sci. Camb. 102:341
11. Reynolds, M.P., R.P. Singh, A. Ibrahim, O.A.A. Ageeb, A. Larque-Saavedra, and J.S. Quick. 1998. Evaluating physiological traits to complimentempirical selection for wheat in warm environments. Euphytica 100:85-9
12. Rosyara, U.R., Vromman, D., Duveiller, E. 2008. Canopy temperature depression as indication of correlative measure of spot blotch resistance and heat stress tolerance in spring wheat. J. Plant Path. 90 :103–107.
13. Rosyara, U. R., S. Subdedi, R. C. Sharma and E. Duveiller.2010.The effect of spot blotch and heat stress in variation of canopy temperature depression, chlorophyll fluorescence and chlorophyll content of hexaploid wheat genotypes. Euphytica Volume 174, Number 3, 377-390
14. Rosyara, U. R., S. Subdedi, R. C. Sharma and E. Duveiller. 2010. Photochemical Efficiency and SPAD Value as Indirect Selection Criteria for Combined Selection of Spot Blotch and Terminal Heat Stress in Wheat. Journal of Phytopathology Volume 158, Issue 11-12, pages 813–821
15. Rosyara, U.R., Vromman, D., Duveiller, E. 2008. Canopy temperature depression as indication of correlative measure of spot blotch resistance and heat stress tolerance in spring wheat. J. Plant Path. 90 :103–107.
16. Hall, A.E., 1992. Breeding for heat tolerance. Plant Breed. Rev. 10, 129–168.
17. Hall, A.E., 2001. Crop Responses to Environment. CRC Press LLC, Boca Raton, Florida
18. Jones, P.D., New, M., Parker, D.E., Mortin, S., Rigor, I.G., 1999. Surface area temperature and its change over the past 150 years. Rev. Geophys. 37, 173–199
19. Schoffl, F., Prandl, R., Reindl, A., 1999. Molecular responses to heat stress. In: Shinozaki, K., Yamaguchi-Shinozaki, K. (Eds.), Molecular Responses to Cold, Drought, Heat and Salt Stress in Higher Plants. R. G. Landes Co.,Austin, Texas, pp. 81–98
20. Ehlers, J.D., Hall, A.E., 1998. Heat tolerance of contrasting cowpea lines in short and long days. Field Crops Res. 55, 11–21.
21. Camejo, D., Jimenez, A., Alarc ´ on, J.J., Torres, W., G ´ omez, J.M., Sevilla, ′F., 2006. Changes in photosynthetic parameters and antioxidant activities following heat-shock treatment in tomato plants. Funct. Plant Biol. 33,177–18

22. Reynolds, M.P., R.P. Singh, A. Ibrahim, O.A.A. Ageeb, A. Larqué-Saavedra, and J.S. Quick.1998. Evaluating physiological traits to complement empirical selection for wheat in warm environments. Euphytica 100:84-95.,

23. Fokar, M., H.T. Nguyen, and A. Blum. 1998. Heat tolerance in spring wheat. I. Genetic variability and heritability of cellular thermotolerance. Euphytica 104:1-8.

24. Rosyara, U. R., S. Subdedi, R. C. Sharma and E. Duveiller. 2010. Photochemical Efficiency and SPAD Value as Indirect Selection Criteria for Combined Selection of Spot Blotch and Terminal Heat Stress in Wheat. Journal of Phytopathology Volume 158, Issue 11-12, pages 813–821, December 2010

25. Rosyara, U. R., S. Subdedi, R. C. Sharma and E. Duveiller.2010.The effect of spot blotch and heat stress in variation of canopy temperature depression, chlorophyll fluorescence and chlorophyll content of hexaploid wheat genotypes. Euphytica Volume 174, Number 3, 377-390

26. Rosyara, U. R., S. Subdedi, R. C. Sharma and E. Duveiller. 2010. Photochemical Efficiency and SPAD Value as Indirect Selection Criteria for Combined Selection of Spot Blotch and Terminal Heat Stress in Wheat. Journal of Phytopathology Volume 158, Issue 11-12, pages 813–821, December 2010

27. Xu W, Subudhi PK, Crasta OR, Rosenow DT, Mullet JE, Nguyen HT. Molecular mapping of QTLs conferring stay-green in grain sorghum (Sorghum bicolor L. Moench). Genome. 2000 Jun;43(3):461-9.

28. Berry J. and O. Bjorkman. 1980. Photosynthetic response and adaptation to temperature in higher plants. Annu. Rev. Plant Physiol. 31:491–543.

29. Physiological approaches to wheat breeding, http://www.fao.org/docrep/006/y4011e/y4011e0a.htm

30. U.R. Rosyara, N.K. Gupta, S. Gupta, R.C. Sharma (2011) Genomics and Phenomics of Chlorophyll Associated Traits in Abiotic Stress Tolerance Breeding In: Hua Le and Elisa Salcedo, Chlorophyll: Structure, Production and Medicinal Uses, N.Y. : Nova Science Publishers

31. Rosyara U. R., Gupta N. K., Gupta S., R.C. Sharma (2012) Breeding for Heat Stress Tolerance in Plants via Molecular and Physiological Manipulation In: Stanislas Josipovic and Elias Ludwig, Heat Stress: Causes, Treatment and Prevention, N.Y. : Nova Science Publishers

32. Rosyara, U.R., Vromman, D., Duveiller, E. 2008. Canopy temperature depression as indication of correlative measure of spot blotch resistance and heat stress tolerance in spring wheat. J. Plant Path. 90 :103–107.

33. Rosyara, U. R., S. Subdedi, R. C. Sharma and E. Duveiller.2010.The effect of spot blotch and heat stress in variation of canopy temperature depression, chlorophyll fluorescence and chlorophyll content of hexaploid wheat genotypes. Euphytica Volume 174, Number 3, 377-390

34. Rosyara, U. R., S. Subdedi, R. C. Sharma and E. Duveiller. 2010. Photochemical Efficiency and SPAD Value as Indirect Selection Criteria for Combined Selection of Spot Blotch and Terminal Heat Stress in Wheat. Journal of Phytopathology Volume 158, Issue 11-12, pages 813–821

35. Rosyara, U.R., Vromman, D., Duveiller, E. 2008. Canopy temperature depression as indication of correlative measure of spot blotch resistance and heat stress tolerance in spring wheat. J. Plant Path. 90 :103–107.

CHAPTER 8

Artificial Production of Doubled Haploids in Plant Breeding

Chapter Contents

INTRODUCTION

A doubled haploid (DH) is a genotype formed when haploid cells undergo chromosome doubling. Artificial production of doubled haploids is important in plant breeding.

Haploid cells are produced from pollen or egg cells or from other cells of the gametophyte, then by induced or spontaneous chromosome doubling, a doubled haploid cell is produced, which can be grown into a doubled haploid plant. If the original plant was diploid, the haploid cells are monoploid, and the term doubled monoploid may be used for the doubled haploids. Haploid organisms derived from tetraploids or hexaploids are sometimes called dihaploids (and the doubled dihaploids are, respectively, tetraploid or hexaploid).

Conventional inbreeding procedures take six generations to achieve approximately complete homozygosity, whereas doubled haploidy achieves it in one generation. Dihaploid plants derived from tetraploid crop plants may be important for breeding programs that involve diploid wild relatives of the crops.

HISTORY

The first report of the haploid plant was published by Blakeslee *et al.* (1922) in *Datura stramonium*. Subsequently, haploids were reported in many other species. Guha and Maheshwari (1964) developed an anther culture technique for the production of haploids in the laboratory. Haploid production by wide crossing was reported in barley (Kasha and Kao, 1970) and tobacco (Burk *et al.*, 1979). Tobacco, rapeseed, and barley are the most responsive species for doubled haploid production. Doubled haploid methodologies have now been applied to over 250 species.

PRODUCTION OF DOUBLED HAPLOIDS

Doubled haploids can be produced *in vivo* or *in vitro*. Haploid embryos are produced *in vivo* by parthenogenesis, pseudogamy, or chromosome elimination after wide crossing. The haploid embryo is rescued, cultured, and chromosome-doubling produces doubled haploids. The *in vitro* methods include gynogenesis (ovary and flower culture) and androgenesis (anther and microspore culture). Androgenesis is the preferred method. Another method of producing the haploids is wide crossing. In barley, haploids can be produced by wide crossing with the related species *Hordeum bulbosum*; fertilization is affected, but during the early stages of seed development the *H. bulbosum* chromosomes are eliminated leaving a haploid embryo. In tobacco (*Nicotiana tabacum*), wide crossing with *Nicotiana africana* is widely used. When *N. africana* is used to pollinate *N. tabacum*, 0.25 to 1.42 percent of the progeny survive and can readily be identified as either F1 hybrids or maternal haploids. Although these percentages appear small, the vast yield of tiny seeds and the early death of most seedlings provide significant numbers of viable hybrids and haploids in relatively small soil containers. This method of interspecific pollination serves as a practical way of producing seed-derived haploids of *N. tabacum*, either as an alternative method or complementary method to anther culture.

GENETICS OF DH POPULATION

In DH method only two types of genotypes occur for a pair of alleles, A and a, with the frequency of ½ AA and ½ aa, while in diploid method three genotypes occur with the frequency of ¼ AA, ½ Aa, ¼ aa. Thus, if AA is desirable genotype, the probability of obtaining this genotype is higher in haploid method than in diploid method. If n loci are segregating, the probability of getting the desirable genotype is (1/2)n by the haploid method and (1/4)n by the diploid method. Hence the efficiency of the haploid method is high when the number of genes concerned is large.

Studies were conducted comparing DH method and other conventional breeding methods and it was concluded that adoption of doubled haploidy does not lead to any bias of genotypes in populations, and random DHs were even found to be compatible to selected line produced by conventional pedigree method.

APPLICATIONS OF DHS PLANT BREEDING

Mapping Quantitative Trait Loci

Most of the economic traits are controlled by genes with small but cumulative effects. Although the potential of DH populations in quantitative genetics has been understood for some time, it was the advent of molecular marker maps that provided the impetus for their use in identifying loci controlling quantitative traits. As the quantitative trait loci (QTL) effects are small and highly influenced by environmental factors, accurate phenotyping with replicated trials is needed. This is possible with doubled haploidy organisms because of their true breeding nature and because they can conveniently be produced in large numbers. Using DH populations, 130 quantitative traits have been mapped in nine crop species. In total, 56 DH populations were used for QTL detection.

Backcross Breeding

In backcross conversion, genes are introgressed from a donor cultivar or related species into a recipient elite line through repeated backcrossing. A problem in this procedure is being able to identify the lines carrying the trait of interest at each generation. The problem is particularly acute if the trait of interest is recessive, as it will be present only in a heterozygous condition after each backcross. The development of molecular markers provides an easier method of selection based on the genotype (marker) rather than the phenotype. Combined with doubled haploidy it becomes more effective. In marker assisted backcross conversion, a recipient parent is crossed with a donor line and the hybrid (F1) backcrossed to the recipient. The resulting generation (BC1) is backcrossed and the process repeated until the desired genotypes are produced. The combination of doubled haploidy and molecular marker provides the short cut. In the back cross generation one itself a genotype with the character of interest can be selected and converted into homozygous doubled haploid genotype. Chen *et al.* (1994) used marker assisted backcross conversion with doubled haploidy of BC1 individuals to select stripe rust resistant lines in barley.

Bulked Segregant Analysis (BSA)

In bulked segregant analysis, a population is screened for a trait of interest and the genotypes at the two extreme ends form two bulks. Then the two bulks are tested for the presence or absence of molecular markers. Since the bulks are supposed to contrast in the alleles that contribute positive and negative effects, any marker polymorphism between the two bulks indicates the linkage between the marker and trait of interest. BSA is dependent on accurate phenotyping and the DH population has particular advantage in that they are true breeding and can be tested repeatedly. DH populations are commonly used in bulked segregant analysis, which is a popular method in marker assisted breeding. This method has been applied mostly to rapeseed and barley.

Genetic Maps

Genetic maps are very important to understand the structure and organization of genomes from which evolution patterns and syntenic relationships between species can be deduced. Genetic maps also provide a framework for the mapping of genes of interest and estimating the magnitude of their effects and aid our understanding of genotype/phenotype associations. DH populations have become standard resources in genetic mapping for species in which DHs are readily available. Doubled haploid populations are ideal for genetic mapping. It is possible to produce a genetic map within two years of the initial cross regardless of the species. Map construction is relatively easy using a DH population derived from a hybrid of two homozygous parents as the expected segregation ratio is simple, *i.e.* 1:1. DH populations have now been used to produce genetic maps of barley, rapeseed, rice, wheat, and pepper. DH populations played a major role in facilitating the generation of the molecular marker maps in eight crop species.

Genetic Studies

Genetic ratios and mutation rates can be read directly from haploid populations. A small doubled haploid (DH) population was used to demonstrate that a dwarfing gene in barley is located chromosome 5H. In another study the segregation of a range of markers has been analyzed in barley.

Genomics

Although QTL analysis has generated a vast amount of information on gene locations and the magnitude of effects on many traits, the identification of the genes involved has remained elusive. This is due to poor resolution of QTL analysis. The solution for this problem would be production of recombinant chromosome substitution line, or stepped aligned recombinant inbred lines. Here, backcrossing is carried out until a desired level of recombination has occurred and genetic markers are used to detect desired recombinant chromosome substitution lines in the target region, which can be fixed by doubled haploidy. In rice, molecular markers have been found to be linked with major genes and QTLs for resistance to rice blast, bacterial blight, and sheath blight in a map produced from DH population.

Elite Crossing

Traditional breeding methods are slow and take 10–15 years for cultivar development. Another disadvantage is inefficiency of selection in early generations because of heterozygosity. These two disadvantages can be over come by DHs, and more elite crosses can be evaluated and selected within less time.

Cultivar Development

Uniformity is a general requirement of cultivated line in most species, which can be easily obtained through DH production. There are various ways in which DHs can be used in cultivar production. The DH lines themselves can be released as cultivars, they may be used as parents in hybrid cultivar production or more indirectly in the creation of breeders lines and in germplasm conservation. Barley has over 100 direct DH cultivars. According to published information there are currently around 300 DH derived cultivars in 12 species worldwide.

The relevance of DHs to plant breeding has increased markedly in recent years owing to the development of protocols for 25 species. Doubled haploidy already plays an important role in hybrid cultivar production of vegetables, and the potential for ornamental production is being vigorously examined. DHs are also being developed in the medicinal herb *Valeriana officinalis* to select lines with high pharmacological activity. Another interesting development is that fertile homozygous DH lines can be produced in species that have self-incompatibility systems.

Advantages of DHs

The ability to produce homozygous lines after a single round recombination saves a lot of time for the plant breeders. Studies conclude that random DH's are comparable to the selected lines in pedigree inbreeding. The other advantages include development of large number of homozygous lines, efficient genetic analysis and development of markers for useful traits in much less time. More specific benefits include the possibility of seed propagation as an alternative to vegetative multiplication in ornamentals, and in species such as trees in which long life cycles and inbreeding depression preclude traditional breeding methods, doubled haploidy provides new alternatives.

Disadvantages of DHs

The main disadvantage with the DH population is that selection cannot be imposed on the population. But in conventional breeding selection can be practiced for several generations: thereby desirable characters can be improved in the population.

In haploids produced from anther culture, it is observed that some plants are aneuploids and some are mixed haploid-diploid types. Another disadvantage associated with the double haploidy is the cost involved in establishing tissue culture and growth facilities. The over-usage of doubled haploidy may reduce genetic variation in breeding germplasm. Hence one has to take several factors into consideration before deploying doubled haploidy in breeding programmes.

Conclusions

Technological advances have now provided DH protocols for most plant genera. The number of species amenable to doubled haploidy has reached a staggering 250 in just a few decades. Response efficiency has also improved with gradual removal of species from recalcitrant category. Hence it will provide greater efficiency of plant breeding.

REFERENCES

1. Jain, S. Mohan, S. K. Sopory, and R. E. Veilleux. 1996. In vitro haploid production in higher plants. Dordrecht: Kluwer Academic Publishers. p.317.
2. Maluszynski et al., 2003.
3. B. Barnabás; B. Obert; G. Kovács (1999). "Colchicine, an efficient genome-doubling agent for maize (Zea mays L.) microspores cultured in anthero". Plant Cell Reports. 18 (10): 858–862. doi:10.1007/s002990050674.
4. International Symposium on Genetic Manipulation in Crops. 1988. Genetic manipulation in crops proceedings of the International Symposium on Genetic Manipulation in Crops, the 3rd International Symposium on Haploidy, the 1st International Symposium on Somatic Cell Genetics in Crops, Beijing, October 1984. Natural resources and the environment series, v. 22. (London: Published for the International Rice Research Institute and Academia Sinica by Cassell Tycooly), p.318.
5. Immonen and Anttila, 1996.
6. Friedt et al., 1986; Winzeler et al., 1987.
7. Ardiel, G.S., Grewal, T.S., Deberdt, P., Rossnagel, B.G., and Scoles, G.J. 2002. Inheritance of resistance to covered smut in barley and development of tightly linked SCAR marker. Theoretical and applied genetics 104:457-464.
8. Blakelsee, A.F., Belling, J., Farhnam, M.E., and Bergner, A.D.1922. A haploid mutant in the Jimson weed, Datura stramonium. Science 55:646-647.
9. Burk, L.G., Gerstel, D.U., and Wernsman, E.A. 1979. Maternal haploids of Nicotiana tabacum L. from seed. Science 206:585.
10. Chen, F.Q., D.Prehn, P.M. Hayes, D.Mulrooney, A. Corey, and H.Vivar. 1994. Mapping genes for resistance to barley stripe rust (Puccinia striiformis f. sp. hordei). Theoretical and Applied Genetics. 88:215-219.
11. Friedt, W., Breun, J., Zuchner, S., and Foroughi-Wehr, B. 1986. Comparative value of androgenetic doubled haploid and conventionally selected spring barley line. Plant Breeding 97:56-63.
12. Guha, S., and Maheswari, S. C. 1964. In vitro production of embryos from anthers of Datura. Nature 204:497.
13. Immonen, S., and H. Anttila. 1996. Success in rye anther culture. Vortr. Pflanzenzuchtg. 35:237-244.
14. Kasha, K. J., and Kao, K. N. 1970. High frequency haploid production in barley (Hordeum vulgare L.). Nature 225: 874-876.
15. Kearsey, M. J. 2002. QTL analysis: Problems and (possible) solutions. p. 45-58. In: M.S. Kang (ed.), Quantitative genetics, genomics and plant breeding. CABI Publ., CAB International.

16. Maluszynski, M.., Kasha K. J., Forster, B.P., and Szarejko, I. 2003. Doubled haploid production in crop plants: A manual. Kluwer Academic Publ., Dordrecht, Boston, London.

17. Paterson, A.H., Deverna, J.W., Lanin, B., and Tanksley, S. 1990. Fine mapping of quantitative trait loci using selected overlapping recombinant chromosomes in an interspecies cross of tomato. Genetics 124:735-741.

18. Schon, C., M. Sanchez,T. Blake, and P.M. Hayes. 1990. Segregation of Mendelian markers in doubled haploid and F2 progeny of barley cross. Hereditas 113:69-72.

19. Thomas, W. T. B., B. Gertson and B.P. Forster. 2003. Doubled haploids in breeding p. 337-350. in :M. Maluszynski, K.J. Kasha, B.P. Forster and I. Szarejko (eds)., Doubled haploid production in crop plants:A Manual. Kluwer Academic Publ., Dordrecht, Boston, London.

20. Thomas, W.T.B., Newton, A.C., Wilson, A., Booth, A., Macaulay, M., and Keith, R. 2000. Development of recombinant chromosome substitution lines: A barley resource. SCRI Annual Report 1999/2000, 99-100.

21. Thomas, W.T.B., Powell, W., and Wood, W. 1984. The chromosomal location of the dwarfing gene present in the spring barley variety Golden Promise. Heredity 53:177-183.

22. Wang, Z., G. Taramino, D.Yang, G. Liu, S.V. Tingey, G.H. Miao, and G.L. Wang. 2001. Rice ESTs with disease-resistance gene or defense-response gene-like sequences mapped to regions containing major resistance genes or QTLs. Molecular Genetics and Genomics. 265:303-310.

23. William, K.J., Taylor, S.P., Bogacki, P., Pallotta, M., Bariana, H.S., and Wallwork, H. 2002. Mapping of the root lesion nematode (Pratylenchus neglectus) resistance gene Rlnn 1 in wheat. Theoretical and applied genetics 104:874-879.

24. Winzeler, H., Schmid, J., and Fried, P.M. 1987. Field performance of androgenetic doubled haploid spring wheat line in comparison with line selected by the pedigree system. Plant breeding 99:41-48.

25. Yi, H.Y., Rufty, R.C., Wernsman, E.A., and Conkling, M.C. 1998. Mapping the root-knot nematode resistance gene (Rk) in tobacco with RAPD markers. Plant Disease 82:1319-1322.

Index

Q

R

S